Li Xinggang

李兴钢 著

李兴钢 2001—2020

LI XINGGANG

浙江摄影出版社

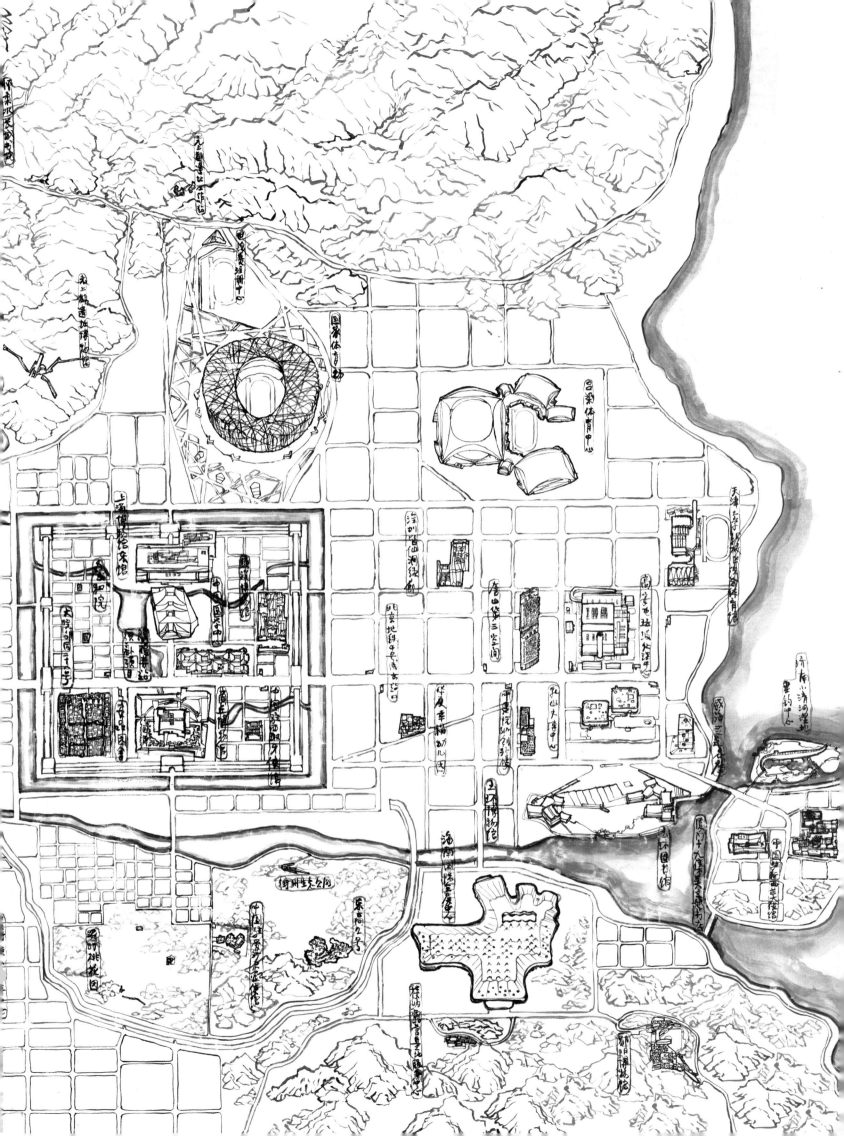

目录
CONTENTS

程泰宁
Cheng Taining

序一
FORWARD 1

我认识兴钢的时间并不长。前些年，除了参加学术活动时偶尔的"惊鸿一瞥"之外，我们第一次见面已是 2012 年在聂兰生先生的家里，恰好我们都去看望聂先生，才有了那次不期而遇。稍感意外的是，虽然是第一次见面，彼此的年龄也完全不在一个量级，但交流时却毫无陌生感。也许，是因为在这次见面之前，我已看过他的不少作品，也听过一些关于他的介绍（聂先生就不止一次在我面前夸赞过他），这些，大概就是我们能像古人所说"倾盖如故"的原因吧。

自那以后，彼此的交流比较多了，这次为了写"序"，我又把他的这本作品集连同另一本《胜景几何论稿》，系统地读了一遍，再次印证了我一直以来的看法：兴钢是一位才华和学识兼具、又正值创作高峰期的建筑师。我们看到他创作的元上都遗址博物馆、海南国际会展中心、天津大学新校区综合体育馆、绩溪博物馆、建川镜鉴博物馆暨汶川地震纪念馆、上海博物馆东馆（竞赛方案）以及南京"瞬时桃花源"、威尼斯纸砖房等项目，多姿多彩，新意迭出，令人过目不忘。而这些风格各不相同的建筑作品，却又贯穿着一条十分明显的思想红线，那就是他对"自然与人工交互"这一理念的强调，以及他在创作中，通过理想的"人工介入"，实现"胜景几何"的不懈探索。

大自然是神奇的，落日、云海、奇峰、怪石常使我们讶异而赞叹。这世间的万事万物，在被抽象提炼和转换以后，常常成为建筑师独特的创意，例如兴钢的乐高一号、二号，这些妙趣横生的参展作品，激起了我们对这种转化的无限遐想。"自然"，永远是建筑师获得创作灵感的永不枯竭的源泉。

但是，正如兴钢所说："自然并非一种天然存在的全然美好。"由于多种原因，包括"人工"的非理性介入，便出现了他所描绘的"满目疮痍"的大漠荒原、杂乱无序的聚落空间，以及肮脏残破的城镇废墟……大自然这些"创伤"的愈合，有赖于积极而理想的人工介入，方能使之成为"现实理想胜境"。这正是人类，也是我们建筑师不可推卸的社会责任。

于是，我们看到了兴钢在他创作的项目中作出的不懈努力和所取得的成果。

元上都遗址博物馆建在离元上都遗址五公里处的乌兰台。为了实现对遗址的"轻介入"，也为了恢复被破坏的自然环境，博物馆的大部分被放置在矿坑内。以广袤的荒原为背景，其地面体量类似烽火堡与堑壕。但博物馆建筑的线性几何造型，自由错动的暗红色墙体，与这片荒原形成了强烈的、但尺度感把握得极好的对比。特别是它的红色矮墙（道路），呼应着等高线由坡底转折而上，直至坡顶敖包，把博物馆建筑与乌兰台"织"成了一个整体。矮墙与建筑所勾勒出的线条刚劲优美，为这片荒原平添了一种生气。兴钢在这个建筑上似是着墨不多，但"人工与自然交互"所产生的效果却堪称完美，体现了建筑师的格局、素养和功力。

I have not known Xinggang for a long time. In the past few years, except for the occasional "glance" when participating in academic activities, our first meeting was in Professor Nie Lansheng's home in 2012. It happened that we all went to visit her and had an unexpected meeting. It was a little surprising that although it was the first time we met, the age gap between us was inconsequential. There was no awkwardness when communicating perhaps because before that meeting, I had seen many of his works, and heard some introductions about him (Professor Nie praised him more than once in front of me). Probably, as the ancients said, some new friends we meet by accident are like old friends with deep friendship.

Since then, we have engaged in more communications. In order to write this Forward, I have systematically read this monograph together with another manuscript, *Essays on Integrated Geometry and Poetic Scenery*. Again, it confirmed my view that Xinggang is an architect with both talent and knowledge, and at the creative peak of his career. The projects of the Museum for Site of Xanadu, the Hainan International Convention & Exhibition Center, the Gymnasium of the New Campus of Tianjin University, the Jixi Museum, the Jianchuan Mirror Museum and Wenchuan Earthquake Memorial, the Shanghai East Museum (proposal), the Instantaneous Peach Garden in Nanjing, the Paper-brick House in Venice and others, represent diverse, new ideas and are unforgettable. These works with different styles have a coherent emphasis on the concept of "interaction between the natural and the artificial" and demonstrate an unremitting exploration of "Integrated Geometry and Poetic Scenery" through ideal "artificial intervention."

Nature is magical, and sunsets, clouds, strange peaks and rocks often surprise us. Everything in this world, after being abstracted and transformed, often becomes unique ideas for architects. These interesting installations, such as the Lego No. 1 and No. 2 projects, present infinite inspirations and revelations to us. "Nature" is always an inexhaustible source for architects to get creative inspiration.

However, as Xinggang has said, nature is not a kind of natural and complete beauty. The irrational, artificial intervention, together with other reasons, gave rise to the desperate scenes he has portrayed, such as a desert wilderness, messy and disorderly settlement spaces, and dirty and ruined towns. The healing of these "traumas" of nature depends on active and ideal artificial interventions, which could produce "realistic ideal spaces". This is precisely an inescapable social responsibility of human beings, including architects.

Accordingly, we see the results of the unremitting efforts and achievements made by Xinggang in the projects he created.

The Museum for Site of Xanadu was built in Wulantai, five kilometers from the ruins of ancient capital of Yuan Dynasty. In order to lightly intervene on the site and also to restore the damaged natural environment, most of the museum's grounds were placed in a pit. With the vast wilderness as its background, the above-ground volume is similar to that of a beacon tower and trench. Additionally, the linear geometric shape of the museum contrasted with the freely

与元上都遗址博物馆不同，绩溪博物馆位于山水环绕的千年古村——华阳镇。面对历史形成的聚落肌理，建筑师以一种和而不同的理念、意到笔不到的设计手法，将博物馆建筑很自然地"安放"在这个古镇之中。"留树为庭""折顶拟山"，打造了一个迥异于传统街巷肌理，却又与已有聚落空间相融合的全新的空间格局。游走在这个空间序列之中，目之所及，变形的屋顶和湖石，黑色的景框和树影婆娑的白墙互相交织，宛如一幅幅现代水墨画，散发着浓浓的人文气息，清新静谧、步移景异，令人神往。

　　海南国际会展中心坐落于大海之滨。第一次看到它的鸟瞰照片时，就感到建筑与自然环境是那么谐和而生动地融合在一起，很喜欢。而且把一个中型会展建筑做得那么灵动，很见功力。这个设计，应该说是功能与形式、建构与意象的高度契合，通过"人工与自然交互"，是表达"胜景几何"理念的一个很好案例。

　　在兴钢的作品中，天津大学新校区综合体育馆显得有些"另类"。仅从外观上看，单元筒拱连接着屋顶和墙面，显示了一种力度和秩序的结构之美。但是在这个项目中，建筑师所追求的似乎并不止于此。仔细阅读过这个设计后，我们会发现：建筑师选择了以结构为切入点，着力打造的则是结构与功能、空间、形式、建造以及采光、通风等细节的有机融合。在取得上述这些问题综合而完善的解决的同时，也取得建筑语言的突破。这体现的是一种整体性思维模式，一种"道法自然"的哲学理念。这样的理念，在兴钢其他很多作品中都有体现：上海博物馆东馆（竞赛方案）、唐山"第三空间"、威海"Hiland·名座"以及中国驻西班牙大使馆改造等项目，都是很好的案例，它们是另一种"人工与自然的交互"，只是这个"自然"，已不是通常的大自然，而是"自然而然"的创作状态了。

　　实际上，兴钢已经提到，"中国文化中的自然，还有另外一种含义，一种自然而然、无为而自成的状态"。这是他对"自然"的另一种解读，一种深层次的哲学解读。在这种解读面前，"建构""范式"以及所谓的"在地性"和"文化性"等等提法就显得刻板而单薄了。因为在创作中，建筑师所面对的物质和精神等诸多因素，是一个相互连接的立体网络，其间的关系具有不确定性和模糊性，诸多因素之间不可排序，也不能采用"功能决定形式"之类的单向思维模式去解决问题。人们所关注的形式，也只有与功能、结构、经济等因素深度结合，才能摆脱语言的苍白和程式化，获得一种独特之美、自然之美。这一观点在兴钢的作品中已经得到印证。因此，可以说，通过理想的"人工介入"，使作品得以"自然生成""浑然天成"，才是创作中"人工与自然交互"的最高境界。

　　细读本书，也让我们进一步了解了兴钢的创作经历。在很多同行心目中，兴钢是一位创作思维活跃而又极具人文气质的建筑师。对此，人们可能会归结于他的先天禀赋，而在这里，我更想强调的却是兴钢自己提到的"自我修炼"。20年来，他"读万卷书""行万里路"、艰苦磨砺、思而后成。兴钢读书涉猎甚广，从段义孚的人文地理学专著到金宇澄的沪语小说，他都能从中撷取自己所需要的滋养；他去过国内外很多地方，领略大自然和建筑经典的美好，心摹手追，画下了大量草图；他多次提到初登景山俯瞰紫禁城时心灵上所产生的巨大震撼；而在佛光寺前，他目睹夕阳、远山、古刹，感受到一种苍凉、悠远的意境并沉浸其中……丰厚的积累和锲而不舍的思考，加之反复的人与境、人与情的深度"交互"、体验和升华，锻造了兴钢敏感而又具有人文气质的艺术感悟力。正是这种感悟力，使他在创作中不断求新求变，而又总是保持了自己的文化品位和追求。这样的"自我修炼"，应该引起我们建筑师，特别是年轻建筑师的思考和借鉴。

　　当前，世界文化格局正在重构，中国现代建筑的发展正处于一个关键时期。作为中国建筑师，我们需要继承传统，也需要借鉴外来文化，但是现实告诉我们，当下中国建筑师最需要的是摆脱"人（古人、洋人）云亦云"的影响，通过传统和外来文化的"创造性转换、创新性发展"，逐步实现中国现代建筑的理论和实践创新，为世界建筑的多元化发展作出自己的贡献。这是挑战，是责任，也是中国现代建筑发展的必经之路。在这样一个大背景下，观照兴钢的创作经历、创作思考和实践，我们可以更清晰地看到他的创作特色：皈依自然、贴近心灵，融汇中西、转换创新。这一创作特色所体现的包容性和创新性，不仅使他的作品在国内外获得了广泛的关注和好评——这在当下价值观比较混乱的情况下十分

shifting dark red wall forms an excellent sense of scale amidst this wasteland. In particular, the red low wall (road) echoes the contour line that turn from the bottom of the slope up to the *aobao* at the top, which weaves together the building and terrace. The low wall and the building's linear shape are strong and beautiful, adding a sense of vitality to this wasteland. The museum was designed modestly, but the effect produced by the "interaction between the natural and the artificial" is perfect, reflecting the architect's vision, accomplishment and skill.

Unlike the Museum for the Site of Xanadu, the Jixi Museum is located in Huayang, Jixi, an ancient town surrounded by mountains and rivers. Faced with the historically formed settlement surroundings, the architect naturally "placed" the museum in this ancient town with unexpected design method. The museum forms a harmonious dialogue with this context, but maintains a subtle distinction. The architect designed courtyards where trees grow, folded the roofs to simulate the outline of mountains and created a brand-new spatial pattern that is very different from the traditional street atmosphere while integrating with the museum grounds within existing settlement space. Wandering in this spatial sequence, one can see deformed roofs and "lake stones", black framing the scene, and the shadows of trees whirling on white walls. This fascinating scene, which looks like a modern ink painting exuding strong humanistic atmosphere, is fresh, quiet, and dynamic.

The Hainan International Convention and Exhibition Center is located on the shore by the sea. When I first saw its aerial photos, I felt that the building and the natural environment were harmoniously and vividly fused together. It is very skillful to make a medium-sized exhibition building so flexible. It should be acknowledged that the design maintains a high consistency of function and form, tectonics and image. The Convention and Exhibition Center is a good example of a building that illustrates the concept of "Integrated Geometry and Poetic Scenery" through "interaction between the natural and the artificial".

Among Xinggang's works, the Gymnasium of the New Campus of Tianjin University appears somewhat special. Viewing the outward appearance only, one can see that the barrel arch units connect the roof and the walls, showcasing structural beauty with a kind of strength and order. But for this project, the architect's exploration is not confined to this. After carefully reading it, we find that the architect chose to use structure as a starting point, focusing on the organic integration of structure, function, space, form, construction, lighting and ventilation. While seeking a comprehensive solution to address these issues, a breakthrough of architectural language has also been achieved. This embodies a holistic mode of thinking and a Taoist philosophy of following nature. This concept is reflected in many other projects including the Shanghai East Museum (proposal), the "Third Space" in Tangshan, the "Hiland · Mingzuo" in Weihai, and the Renovation of Chinese Embassy in Spain. They are also good examples of the "interaction between the natural and the artificial". This "nature" is no longer the usual nature, but a "natural" free state of creation.

In fact, as Xinggang has already mentioned, "nature" in Chinese culture has another meaning, a condition of being natural, of doing nothing and self-formation. This is an alternative understanding of nature and a deeper philosophical interpretation. Compared with this interpretation, the discourses of "tectonics", "paradigm", "locality" and "culture" appear to be inflexible and weak. In the design process, many factors such as the material and the spiritual faced by architect form a three-dimensionally connected network, and the relationship between them is uncertain and vague. Many factors cannot be sorted and problems cannot be addressed with the singular thought paradigm of "form following function". The form that people pay attention to can get rid of the mundane and stylization of language and gains unique natural beauty only through deep integration with function, structure, economy and other factors. This view has been confirmed in Xinggang's work. Therefore, it can be said that work can be naturally generated through the ideal "artificial intervention". This is the highest level of the "interaction between the natural and the artificial" in architectural creation.

Reading this book carefully also gives us a better understanding of Xinggang's design experience. In the eyes of his peers, Xinggang is a creative architect with a strong humane temperament. In this regard, people may attribute it to his innate talent. But what I want to emphasize here is the self-cultivation that he has credited. In the past 20 years, he has read extensively, travelled widely, and worked diligently. Xinggang has a wide range of reading interests, from Yi-fu Tuan's monograph on human geography to Jin Yucheng's Shanghainese novels, from which he can extract the nourishment he needs. He has visited many places at home and abroad to appreciate the beauty of nature and architectural classics. Impressed by those landscapes, he drew a lot of sketches. He repeatedly mentioned the great shock in his mind when he first climbed Jingshan Mountain and saw the Forbidden City from this unique vantage point. In front of the Foguang

不易——更为重要的是，这一创作特色的形成，是他 30 年来，不断摆脱各种思潮的羁绊，不闻风起舞，也不为时尚所惑，独立思考、艰苦探索的结果。他，走出了一条他自己的路，而这条路，是符合世界文化发展的大方向的。他的创作"路向"和这一"路向"所引发的生机勃勃、佳作迭出的创作状态，令人心动，值得我们深思。

"长风破浪会有时，直挂云帆济沧海"，衷心祝愿兴钢在世界建筑的"大海"中踏浪凌波，勇立潮头。

是为序。

大疫居家之时

Temple, he witnessed the sunset, gazed at distant mountains and ancient temples. He felt a desolate and distant artistic conception and was immersed in it. The rich accumulation and perseverance of thinking, combined with the experience and sublimation of the repeated interaction between people and environment, and between people and emotions, forged Xinggang's sensitive and humanistic artistic perception. It is this sensibility that makes him constantly seek innovation and change in his practice, while always maintaining his cultural taste and pursuits. Our architects, especially the younger generation should think and learn from this self-cultivation.

At present, the global cultural landscape is being reconstructed, and the development of modern Chinese architecture is in a critical period. As Chinese architects, we need to inherit traditions and learn from foreign cultures, but reality tells us that what we need most today is not to parrot what others say (both ancients and foreigners), but to transform and develop traditional and foreign cultures, both creatively and innovatively. Only in this way can we gradually achieve theoretical and practical innovation in modern Chinese architecture and make our own contribution to the diversified development of global architecture. This is a challenge, a responsibility, and a necessary process for the evolution of modern Chinese architecture. From such a large background to observe Xinggang's creative experience, thinking and practice, we can see more clearly his design characteristics: paying tribute to nature, following the mind, combining both Chinese and Western culture, and transforming them through innovation. The inclusiveness and innovation embodied in these characteristics contributed to the widespread attention and praise his work has received at home and abroad. Given the confusing values existing in current situation, this recognition is not easy. More importantly, the formation of these characteristics is the result of his independent thinking, self exploration and continuous efforts to create, unfettered by various trends of thought in the past 30 years. He neither chased the trend, nor was confused by fashion, but walked his own path. This path is in line with the general direction of cultural development in the world. His design "direction" and the dynamic and creative state of work appearing along this direction are exciting and deserve our deep consideration.

"Someday, with my sail piercing the clouds, I will mount the wind, break the waves, and traverse the vast, rolling sea." I sincerely wish Xinggang to ride the waves in the sea of global architecture and brave the tide.

Cheng Taining
February 2020
Written at home during the pandemic

Cheng Taining is an architect and Academician of Chinese Academy of Engineering.

刘家琨

Liu Jiakun

序二
FORWARD 2

兴钢和我的友情始于2003年。那年夏天（至少气候对我来说是夏天），我们随亚建协中国代表团一起去了孟加拉国。团友大多是各大院的老总，只有我是民营小作坊的，我自己觉得有点疏离。兴钢在团里最年轻，而我不会英语，于是他成了我的谈伴兼翻译。

我们应该都是第一次去南亚次大陆：浮莲铺展的洪泛平原，苍翠葱绿的丛林，自生自灭的众生，五彩斑斓的贫困……当然，还有路易斯·康永恒遗作中的那些光明和幽暗——朝圣之旅使兴钢年轻的脸上光芒四射。

达卡司机慢吞吞的南亚时间观，把原本精心计划的印度洋观夕阳之旅眼睁睁拖成了暗夜听海，而一车人的急躁与失望，把司机搞得更是一脸困惑：不就是没赶上吗？没赶上就是没赶上，夕阳明天还有的啊。这种节奏感，既适合兴钢说出他的烦恼，也适合我说出我的安慰。兴钢觉得自己已经做了很多设计，却很少建成，不知道是什么原因，心里有点焦急。而我劝他不要着急。首先，他当时的年龄比我开始做建筑时的年龄还小，而知识储备和技术训练却比我扎实太多了。此外，我相信时令节气和"一万小时定律"，每个人的路上都有很多必须要填的坑，我不清楚兴钢的坑在哪里，但一看就是优等生的李兴钢，一定也有他的坑必须要填。不着急，急也没用，夕阳明天还有的啊。

几年以后，"鸟巢"使兴钢声名大噪。是的，"鸟巢"是个国际合作项目，而兴钢能够负责把构想做到成功落地，当然成就显著，其中的努力和艰辛不难想象——或者难以想象。羡慕嫉妒之余，有时候我觉得这也许就是兴钢的大坑：基本上，好多才华初绽的青年建筑师，自从参与过国家重大项目后，顺理成章一路青云，从此走上了组织化、经营化乃至管理化的领导岗位，由此也就疏离了初学建筑时的那个痴迷于创作乐趣的自我，这当然也没有什么不好，应该说更好，只不过是我这个人希望兴钢这个人可能有其他选择。

兴钢果然不同："鸟巢"这么大个事，他居然在自己的作品集里提都不提！这本作品集是二十年辛勤探索的成果——几十个建成项目！和大多数作品集相似，这些作品也是按编年排序的。岁岁年年大不相同，大院又忙得厉害，身为总建筑师，更是需要三头六臂，需要世事练达，甚至需要牺牲自我，因此作品集常常呈现嫁鸡随鸡、遇佛杀佛的状态也不足为奇。而兴钢如此之多的建成作品，却有着内在的统一性：在"理想实践"的主旨下，这些不同状况下的作品，无论大小和类型，都有着对传统城市和园林旨趣的明明暗暗的借鉴和分门别类的试验，显然这已经成为兴钢一以贯之的方法。"胜景几何"，试图把传统意境与当代技术糅合得骨肉匀停，是一件难度极高的事。状态佳的时候，兴钢会接近理想：天津大学新校区综合体育馆浑厚拙朴，绩溪博物馆明净深远，元上都遗址博物馆苍凉遒劲，建川镜鉴博物馆暨汶川地震博物馆粗粝迷

The friendship between Xinggang and me started in 2003. That summer (at least the climate was like summer for me), we went to Bangladesh with the Chinese delegation of the Architects Regional Council Asia (ARCASIA). Most of the delegates were the leaders of state-owned design institutes. Only I was from a small private design firm. I felt a little alienated. Xinggang was the youngest in the group, and I did not speak English, so he became my partner and interpreter.

This was our first experience in the South Asian subcontinent: the flooded plains coverd with floating lotuses, the lush green jungle, but also the neglected people and the poverty... Of course, amid the light and darkness emerged the eternal legacy of Louis Kahn. This pilgrimage made Xinggang's young face shine.

The Dhaka driver's leisure sense of time turned the originally planned sunset view of the Indian Ocean into a journey of listening to the sea through the dark night. The delegates' irritability and disappointment made the driver much more confused: didn't you just miss the chance to see the sea? If you do not look, you will miss it. Besides, there will be another sunset tomorrow. Nonetheless, the pace of this excursion was suitable both for Xinggang to express his concerns and for me to express my advice. Xinggang felt that he had completed a lot of designs, but they were rarely built. He did not know what the reason was, so he felt anxious. However, I advised him not to worry. First of all, he was younger than I was when I started my career as an architect, but the knowledge he had already gathered and the technical training he gained were far more solid than mine. In addition, I believe solar terms and the "10,000-hour rule". Everyone has weaknesses to be improved during a career. I do not see a weakness, but for an overachiever, Xinggang continually seeks improvement. Do not worry, as it is useless to be worried. There will be sunset tomorrow anyway.

A few years later, the "Bird's Nest" (National Stadium) project made Xinggang well-known. He was responsible for the successful implementation of this international collaborative project. Indeed, his achievements are remarkable. It is not difficult to imagine his effort and hardship, but perhaps they are unimaginable. With envy and jealousy, sometimes I think that this project may be an obstacle for Xinggang. Many talented young architects, after participating in major national projects, are relegated to administrative roles responsible for organization, operation or management, thus neglecting their passion for creativity that drove them to architecture. Of course, there is nothing wrong with it. It should be said that this is a better result. However, it is just my hope that Xinggang may have other options.

Xinggang is unique. The "Bird's Nest" , a world-renowned project, is not mentioned in this monograph! This collection of works begins with a drawing full of bloody tears during his student period and includes the results of 20 to 30 years of hard work—with dozens of built projects. Similar to most architectural monographs, these works are also sorted chronologically. Every year is different and the atelier is always busy. As its chief architect, he must demonstrate talent, sophistication, and self-sacrifice. Therefore, it is not surprising that the monograph often shows both compromise and dedication. But Xinggang's numerous completed works have an inherent unity. Under the theme of "ideal practice", he

幻……状态一般的时候，兴钢也会略显生涩，一般来说是稍稍做过，而不是没有做够，但事无巨细都能够一直保持这种拼命三郎式的执着和投入，真的很不容易！不言而喻，这是兴钢的素质决定、个性使然，此外，也许这真的就是崔愷主推的在大院中设立个人工作室这种"大船中有小船"的策略所取得的一项综合成果。

耐心、专注、勤奋、扎实的功底和挫折失败的积累，终于使兴钢渐入佳境。一大堆国家级重大项目，加上兴钢这种烟酒不沾刻苦钻研、饭都少吃苛求极致的性格，使他这些年来把建筑学的十八般武艺都掌握了，换句话说，他至少已经取得了技术自由。

作品集最后，居然是一个由工地脚手架和农用遮阳网搭建的临时综合装置——"瞬时桃花源"。在一系列重量级作品之后以"轻"作为压轴，有点儿意味深长，就像是转入舒缓随心的预示，接下来应该是追求心性自由了吧，我觉得。

"神造万物，各按其时成为美好。"兴钢已经很好了，而兴钢依然年轻。

二〇二〇年四月

created these works under different conditions, regardless of size and type. His designs show a clear foundation that draws lessons from traditional cities and gardens design, while still innovating within their respective categories. Clearly, this method has become a consistent method for him. His exploration of "Integrated Geometry and Poetic Scenery", which attempts to properly combine traditional artistic conception with contemporary technology, is extremely difficult. When his imagination flows, Xinggang nearly reaches his ideal: the Gymnasium of the New Campus of Tianjin University is sturdy and unpretentious; the Jixi Museum is bright and far-reaching; the Museum for Site of Xanadu is bleak and energetic; the Jianchuan Mirror Museum and Wenchuan Earthquake Memorial is rough and illusive. When his state is more conventional, Xinggang is also slightly immature, with his work overdesigned, rather than less. However, it is difficult to sustain such a level of obsession and dedication to all matters, whether important or trivial. It goes without saying that this reflects his professionalism and personality. In addition, perhaps this is really one of the comprehensive results of the strategies that Cui Kai promoted to set up personal studios within the design institute.

Patience, concentration, diligence, solid skills and the accumulation of setbacks and failures gradually made Xinggang better. He pursues perfection—he works hard, does not smoke, does not drink, and eats less. After designing a large number of national major projects in recent years, he has mastered the "eighteen martial arts" of architecture. In other words, he has at least achieved freedom at the technical level.

The end of this book features a temporary, integrated installation constructed with scaffolding and shading nets used in agriculture—the Instantaneous Peach Garden. It is meaningful that this "light" project follows a series of "substantial" works. It seems to suggest a transition to a soothing period of calm and freedom. The next step is to pursue freedom of mind and nature, I think.

"God makes all things beautiful according to their time." Xinggang is already very good, and still young.

Liu Jiakun
April 2020

Liu Jiakun is the founder and principal of Jiakun Architects in Chengdu, China.

李兴钢
Li Xinggang

胜景几何：
与"自然"交互的建筑
INTEGRATED GEOMETRY AND POETIC SCENERY: ARCHITECTURE INTERACTING WITH "NATURE"

这部作品集记录了我自 2001 年以来刚好 20 年的代表性工作。列在所有建筑作品之后的南京"瞬时桃花源"（2015），是一个存留时间不足一月的建筑装置，并非严格意义上的建筑作品，之所以收录于此，是因为这个作品的特殊意义。它是经历多年实践和思考之后逐渐形成的"胜景几何"思考的一次综合性在地建筑实验，朴素地呈现"现实理想空间范式"的初胚、内核和某种理想"标本"，是之前和后来许多真正建筑实践的起点和延伸。

阿尔伯蒂（Leon Battista Alberti，1404—1472）在《论建筑》中说："我所称之为建筑师的人，从完美的艺术与技巧的角度来说，是通过思考与发明，既能够设计，也能够实施的人；是对于（建筑）工作过程中的所有部分都了如指掌的人；是通过对巨大重物的移动，对体量的叠加与联结，能够创造出与人的心灵相贯通的伟大的美的人。"他对建筑师的这一定义使我深深信服，自 1991 年由天津大学建筑系学习毕业到北京成为一名职业建筑师开始，我就将此作为自我修炼的至高目标：建筑师应该是工匠、人文知识分子、艺术和科学工作者合而为一的人，应该具备运用建筑语言表达自己思想的能力，利用各种资源和手段建造实现自己作品的能力和通过建筑对社会、文化、时代进行批判性思考的能力。

记得天津大学建筑系的邹德侬先生曾在毕业纪念册上留言："盖起一大批好房子来。"如今许多年过去，这些已落成于大地的房子在面前铺展开来，我仍然能在瞬间回忆起每个项目中投入的激情、产生的困扰和各自不同的艰辛历程，以及那些与每个建筑相关的人和故事。我不禁自问：它们都算是"好房子"吗？

什么样的房子才算是好房子？这是一个需要被不断追问的问题。

我曾多次提及大学时登临景山万春亭俯瞰故宫的经历：一种直击心灵深处的强烈震撼，使我意识到在种种有关建筑的定义和案例之外，有一种伟大而独特的建筑和城市，我深信其中一定蛰伏着高妙的"设计"或者"意匠"，并且是一种不随时间而消逝、有长久价值的"传统"，它基于一种特定的文化和哲学，产生强大的精神性力量，触动人心，其中的奥秘所在值得探寻。

在 2001 年以前的十年，我的实践聚焦于传统中国"城市"与"建筑"及其同构、复合关系的思考和当代转化，诸如北京兴涛学校和北京兴涛会馆、北京大兴区文化中心等，实际上是我的毕业设计"华人学者聚会中心（1991）"在工程项目中的延伸实验和实践。后来的商丘博物馆（2015）是第一个把这种意图完整落地实施的项目。

在 2001 年完成的北京兴涛接待展示中心中，我第一次借鉴园林的元素并尝试将其引入当代建筑设计，试图将建筑的商业特征与中国传统园林的空间体验和东方意味融合在一起，这也是有关中国园林与当代建筑关系之兴趣、思考和实

This monograph records 20 years of my representative work since 2001. Presented after all the projects of architectural work, the Nanjing Instantaneous Peach Garden designed and realized in 2015 was an architectural installation that stood for less than a month, but not strictly an architectural work. I included this project in this book precisely because it is a comprehensive architectural experiment conceived from the idea of "Integrated Geometry and Poetic Scenery". I gradually formed after many years of practice and reflection. It plainly presented the first embryo, core and some kind of ideal "specimen" of "realistic ideal space paradigm", which is the starting point and extension of many real architectural practices before and after.

In his book *De re aedificatoria* (On the Art of Building), Leon Battista Alberti (1404—1472) wrote, "The person I call an architect is, in terms of art and technique, a person who can both design and build through thought and invention; a person who is familiar with the entire (architectural) work process; a person who can create a great beauty by positioning materials with one's imagination to connect with the masses." This definition of architect resonates with me. Since my graduation from Tianjin University in 1991 and my current status as a professional architect in Beijing, this is my aspiration: an architect should be a person who is able to combine the abilities of artisans, intellectuals, artists, and engineers, and use an architectural vocabulary to express ideas. An architect uses these resources as a means to create structures that critically reflect upon the contemporaneous culture.

I keep in mind what Zou Denong, a professor from the Department of Architecture at Tianjin University, wrote on yearbook: "Build many good buildings." Many years later, I consider the images of the many buildings spread before me. I can still remember the passion, the distress, and different hardships invested in each project, and people and stories related to each building. I cannot help, but wonder: are they all considered "good buildings"? What is a "good building"? This is a question that I must continuously consider.

I have mentioned many times my experience of visiting the Jingshan Wanchun Pavilion and overlooking the Forbidden City when I was a university student: a shock that penetrated my soul, making me realize that it is a kind of great and unique architecture beyond all the definitions and cases. I was deeply convinced that there must be a subtle "design" or "artistic conception" dormant in it, and a "tradition" with long-lasting value that does not fade over time. Derived from a specific culture and philosophy, it generates a palpable spiritual power that touches peoples' hearts, while harboring a hidden secret worth discovering.

In the decade before 2001, my practice focused on the examination of the contemporary transformation of traditional Chinese cities and their isomorphic and compositional relationship with architecture. The projects such as the Beijing Xingtao School, the Beijing Xingtao Club, and the Beijing Daxing District Cultural Center, are actually extended experiments and practices of the Chinese Scholars' Gathering Center, which was a schematic proposal submitted in 1991 for my graduation design. Later, the Shangqiu Museum(2015) was the first project to fully implement this intention.

In the Beijing Xingtao Reception and Exhibition Center completed in 2001, I, for the first time, tried to introduce the elements of gardens and used them as reference in contemporary architectural design. My aim was to integrate the building's commercial features with the spatial experience and impression of traditional Chinese gardens. This was the beginning of my interest in, thinking on and practice of the relationship between Chinese gardens and contemporary architecture. The Jianchuan Mirror Museum and Wenchuan Earthquake Memorial projects, the Renovation of No. B-59-1 Fuxing Road in Beijing, and even the Museum for Site of Xanadu built in the following years are the tangible results of my exploration of the aforementioned relationship.

From 2003 to 2008, cooperating with Swiss architects Herzog and de Meuron to design and build the National Stadium (the main stadium of the 2008 Beijing Olympic Games), the "Bird's Nest" enabled me to experience and witness this great building's monumental process from conception to reality. I began to develop a strong interest in the interaction between ontological architectural elements (such as structure, form, space, and construction based on geometric logic), with their evolution into a building that presents its essence within a people-oriented spatial field. At the same time, the process also made me realize that architecture is able to capture and transform landscape. A structure can also interact with its environment to inspire a temporary vibrant space through this artificial intervention into "urban nature."

The 2008 Paper-brick House in Venice was a commissioned exhibition work for the China Pavilion of the 11th Venice International Architecture Biennale. Designed and constructed after the Wenchuan Earthquake in May of that year, it was a light building using special materials and construction methods to gently respond to nature, rather confronting it. The

践的开始，之后几年的建川镜鉴博物馆暨汶川地震纪念馆、北京复兴路乙59—1号改造，乃至元上都遗址博物馆等，都是在这一主题下的进一步实践。

2003—2008年，与瑞士建筑师赫尔佐格和德默隆合作设计建造国家体育场（2008年北京奥运会主体育场）——"鸟巢"，使我亲历和见证这一历史性的伟大建筑从构思变为现实的艰难过程，并开始对结构、形式、空间、建造等建筑本体元素基于几何逻辑的互动而不断衍化为呈现其本质的建筑及其以人为主体的空间场域产生浓厚的兴趣，同时，还使我意识到建筑对景观的捕获与加工——结构也可以与自然互动，形成人工对"城市自然"的巧妙介入和因缘际会的生动场所。

2008年的威尼斯纸砖房，是在当年五月汶川大地震后设计建造的第11届威尼斯国际建筑双年展中国馆应邀参展作品，采用特殊的材料，以柔和的方式应对自然的轻型建筑，而非以抵抗的方式应对自然的重型建筑，通过这种方式，我重新思考大自然的威力和建筑中人与自然相处的方式，并思考"普通建筑"所对应的广袤普芄的大众日常生活。

2009年开始的唐山"第三空间"综合体，开启我对传统及当代"聚落"的研究和思考。在普遍平庸的城市环境中，探索了居住空间的个性化特征，标准层楼板层叠交错，赋予了居住生活多样的体验性和视觉化特征，在城市中心高密度环境中实现"理想居住"；同时，建筑自身成为城市中的新景观。

1976年7月28日凌晨唐山大地震发生时，七岁的我并不知道发生了什么事情，但之后余震发生时，自己无论如何都站不稳，好像大地要被掀起来，人在那时就跟天地之间脆弱渺小的动物一样，原来坚固无比的家和房子成了危险与脆弱的物体，不再能庇护我们，心中油然升起巨大而莫名的恐惧感。这样的印象深深透入骨髓，终生难忘，使我本能地意识到：在大自然面前，人是如此渺小，自然是无法被人"征服"的，人与自然之间，不应该是"征服"与"被征服"的关系。大地震以特殊的方式提醒人类，需要充分认识自然的威力和人与自然相处的方式，否则，让人们最有安全感的家园，将在某一时刻使他们迫不及待地逃离。可能就是这样的特殊个体经验，使我逐渐形成了对于"自然"以及人与自然之关系的一种本能性的、根深蒂固的认识。

儿时家乡的那些特别的场景——冬日的田野、萧瑟的树木、干涸的河流、静默的村庄，都给我特别的触动，我喜欢这样一种与自然相关的诗意场景，而延伸到一些文学、艺术作品所表达的类似意境，我自然会有同感与通感。多年以后开启对中国园林的兴趣、考察和研究，从中感受到一种久已逝去却倍感契合的生活哲学活生生地存在其间，由自然物与人造物在限定空间中的相互巧妙响应而产生，具有一种令人惊讶的当代性，我被深深打动并沉醉于这个奇妙的由人类自造的完整世界。这样的个体经验，使我感到可以通过营造而在自然与人之间产生一种深具魅力的交互关系，对我而言，这是一个深具启发性的传统，它可以超越时间，留存于当代，也可以延伸向未来。

1990年"爬景山瞰故宫"的经历和场景在我后来的阅历中不断出现：无论是登高俯瞰蔚县破败的小村庄，抵临观音阁观音像前放眼眺望蓟州古城，或者站在佛光寺东大殿平台遥望夕阳远山，那种人、景、物交相辉映的胜景，都让我深深触动。而这种登高远眺获得的特别感受，也影响了我后来的设计方法：设置立体的登高路径，建立新维度的视觉情境交互，成为我常用的营造建筑诗意的手法之一。为什么经历中的那些场景比纯粹的自然景观更丰富、感人和震撼？我想就是因为那是一种人工与自然高度交相辉映的结果。因为人工介入和影响的自然带着人类生活的痕迹和记忆，所以更容易被人感知并产生诗意的通感。

当多年以来的思考和研究逐步涵盖城市、建筑、园林、聚落之后，我越来越认识到一种存在于四者之中的"一体性"：它们不过是中国传统营造体系的不同方面与不同方式的存在和呈现，它们基于相同的生活哲学——对建筑与自然更为紧密互动、相互依存、共生共长之关系的格外关注，一种与特定人群的生活理想密切关联之状态的营造。中国建筑师对于传统之于现当代建筑传承及启示的探索，可谓久已有之，有看重立面风格形式者，有着力平面空间构成者，而我则将营

Paper-brick House rethought of the power of nature and the way people and nature exist amidst architecture, and considers how ordinary architecture can respond to ordinary people.

The Tangshan "Third Space" Complex project, which began in 2009, enabled me to start research and contemplate traditional and contemporary "settlements". I explored the personalized living space within the generally mundane urban environment. The complex's layered and interlaced standard floors result in a varied visual and residential experience. My goal was producing "ideal living" in the high-density environment of the city center, while simultaneously, creating a building that by itself becomes a new urban landscape.

When the Tangshan earthquake happened in the early morning of 28 July 1976, I was seven years old and did not know what happened. But when the aftershocks of the earthquake occurred, I felt anxious, as if the earth was about to be overthrown. People at that moment were like fragile, small animals helpless within nature and the world. Given the fact that existing "sturdy" homes and houses became very dangerous and fragile objects that no longer protected us, a considerable and inexplicable fear rose in my heart. Such an impression penetrated deeply through my bones into the marrow and became unforgettable throughout life. I instinctively realized that man, in the face of nature, is so small that nature cannot be "conquered" by man. The relationship between man and nature should not be the relationship between the "conquerer" and the "conquered". The earthquake reminds, in a profound way, that human beings need to fully understand the power of nature and how to get along with nature. Otherwise, people would be eager to flee from their secure homes. It may be this particular individual experience that made me gradually form an instinctive, deep awareness of "nature" and the relationship between man and nature.

The special scenes of my childhood hometown—the winter fields, the sorrowful trees, the dry rivers, and the quiet villages—all left an indelible mark on me. I appreciate poetic scenes related to nature. My appreciation extends to similar artistic reflections expressed in other works of literature and art, with which I share a kinship. Many years later, I started my interest, investigation and research on Chinese gardens. I felt a long-lost, but familiar life philosophy living in them. This unexpected life philosophy emerged from the interaction between natural and artificial objects in a limited space. I was deeply moved by and immersed myself into this wonderfully self-contained world created by humanity. Such a personal experience allowed me to believe that it is possible to create a symbiotic relationship between nature and people through design and construction. For me, the construction pattern of Chinese gardens is a deeply inspiring tradition that transcends time, which must be preserved from now into the future.

Later in my journeys, the evocation of climbing the Jingshan Wanchun Pavilion and looking beyond at the Forbidden City continued to appear: climbing and overlooking the dilapidated, small villages of Yu County, looking at the ancient city of Jizhou while arriving in front of the Guanyin statue in the upper floors of the Guanyin Pavilion, and standing on the platform of the East Hall of Foguang Temple and looking out the setting sun, the mountains in the distance. All these poetic scenes of the interplay between people, scenery, and structures touched me. The special feeling I got from climbing and peering into the horizon also affected my later design methods. Setting up an ascending path to establish a new perspective of visual interaction with one's surroundings has become one of my methods of creating architectural harmony. Why are these experiences and perspectives richer, more touching, and thought provoking than undisturbed natural landscapes? I think that it is a result of the complex interaction between the artificial and the natural. As the natural and artificial intersect, and evoke human experience, it is easier for people to recognize the poetic synergy architecture can produce.

After years of thought and research, gradually exploring cities, architecture, gardens, and settlements, I have increasingly realized that there is a "oneness" among the four. Although existing and presented in different environmental contexts, traditional Chinese design and construction are based on the same philosophy, life, and thought, that is, an extra attention to the closer, interdependent, and symbiotic relationship between architecture and nature, a kind of design and construction closely related to the life ideals of specific people. Chinese architects have long explored the issues of tradition and its inspiration to modern and contemporary architecture, either from the perspective of facade, style and form, or from plan and space composition. However, I tend to consider the creation of interaction between artificiality and nature as a point of departure, to link design thinking and practice with life philosophy and cultural spirit behind urban and architectural phenomena, and to integrate them closely with the complex and diverse human era and reality. This is the core of the architectural concept "Integrated Geometry and Poetic Scenery".

造人工与自然的交互关系作为切入点，把设计思考和实践跟城市及建筑现象背后的生活哲学与文化精神关联起来，并与复杂多元的时代和现实紧密结合起来，这也即是"胜景几何"建筑理念的核心内容。

随着生活和实践阅历的增加，我对"自然"的思考和理解也愈加拓展为一种更为广义的"自然"，可分为原生自然和人工自然：

原生自然／荒野自然，例如山川、草木、土石、天海、风云雨雾、动物，人也是原生自然的一部分；

人工自然／文明自然，例如聚落、园寺、城池、遗迹、历史建筑或城市，甚至承载某种人类生活记忆的当代建造，也可成为人工自然。

承载当代人的生活状态的多样现实造就了多样的"自然"，既有原生自然／荒野自然，也有人工自然／文明自然。例如我们的建筑项目所在的各种建成环境：绩溪古镇和天津大学新校区、北京旧城大院胡同和唐山震后新城区、首钢工业遗址园区和延庆世园小镇等。

梁思成先生在1943年完成的《中国建筑史》绪论中即已提出："建筑显著特征之所以形成，有两因素：有属于实物结构技术上之取法及发展者，有缘于环境思想之趋向者。……政治、宗法、风俗、礼仪、佛道、风水等中国思想精神之寄托于建筑平面之分布上者，固尤甚于其他单位构成之因素也。"依我的理解，真正的"环境思想"，是将梁思成所谓"实物结构技术"与"自然"环境高度交互结合的结果，而其中"政治、宗法、风俗、礼仪、佛道、风水等"代表"环境思想之趋向者"的"中国思想精神之寄托"的，即是不可或缺的广义"自然"。

在我的观察和认识中，"人工"与"自然"并非一成不变，而是在特定条件下能够相互转化的，人工之物可以具有某种自然性，自然之物也可由人工和纯自然元素来混合"制造"。人工与自然之界限的模糊化，可以使两者相互衍生与转化为"自然化的人工"和"人工化的自然"；理想的"人工"是介入自然的人工，最美的"自然"乃是被人工所介入的自然。

我们所孜孜探求的"理想建筑"，应是能让人获得诗意的精神体验的建筑。建筑抵达诗意的方式很多：建筑的纪念性和建构性可以带来仪式感和物质性的诗意；古人笔下的桃花源、海杜克纸上建筑中所描绘的令人向往的生活方式可以带来"理想化"的诗意，而我则选择"建筑与自然交互"这样的途径来获得诗意。"交互"，是指两个相反相成事物的互动和关联。

"胜景几何"，即是营造与"自然"交互的建筑。当代现实中的多样"自然"，需要相应的"人工"策略与之相互动。我们在实践中需要面对的"自然"如此多元化，不同地域、环境、历史中的场地，呈现出"自然"的丰富差异，因此对于场地内外的"自然"之状态的观察、感知、研究、判断至关重要。如何看待不同的"自然"，用怎样的建筑策略与"自然"交互？2010年之后至今的10年，我们的建筑实践更加聚焦于此，呈现出多样的"自然"及其与建筑的各样交互状态，为使用者营造各种不同现实中的当代理想生活空间。

绩溪博物馆和天津大学新校区综合体育馆所在的场地，是现实中国当代城市特征的两极案例。绩溪是一个典型的既拥有山水环抱的"原生自然"又拥有历史沉淀的"人工自然"的千年古镇，我称之为"丰饶自然"；而天津大学新校区，则是平地而起的新城中一片没有任何场所感和归属感的空白之地，我称之为"空白自然"。对于绩溪博物馆所面对的"丰饶自然"，采用了"因借与经营"的方式，使建筑的内外空间、结构、形式与山水树木等"原生自然"和古镇民居等"人工自然"交互作用，让人们体验到超越时空与生命的诗意情境；而对于天津大学新校区综合体育馆所面对的"空白自然"，则需要自造一种由结构"聚落"而形成的"人工自然"，利用内部空间中人的运动这一"特殊自然"，并使建筑与之交互作用，填充现实的空白，营造出动人的空间归属感和场所感。

北京大院胡同28号院和唐山"第三空间"所在的场地，是当代中国城市居住空间模式的两极案例。大院胡同是典

As my personal and professional life evolves, my thinking and understanding of "nature" has expanded from "nature" , into primordial nature and artificial nature. Primordial/wild nature are, for example, mountains, rivers, plants, earth, rocks, sky, seas, wind, clouds, rain, and fog. Animals, including human beings, can also be considered as parts of the primordial nature. Artificial/civilized nature, for example, include settlements, gardens, temples, city-walls and moats, relics and historic buildings or cities. Contemporary constructions derived from human experience can also transform into artificial nature.

The diverse realities of contemporary human life result in the merging of both primordial/wild nature and artificial/civilized nature. For example, the various built environments where our architectural projects are located reveal these dual identities: the Jixi ancient town, the new campus of Tianjin University, the Dayuan Hutong in historic Beijing, the post-earthquake new town in Tangshan City, the Shougang industrial heritage park and the town of the World Horticultural Exposition in Yanqing.

In the introduction of *History of Chinese Architecture*, which was completed in 1943, Liang Sicheng wrote: "There are two factors in the formation of the distinctive features in architecture: the method and development of the technique and formalism, and environmental influence...Politics, a patriarchal clan system, customs, etiquette, Buddhism, Taoism, Feng shui and other Chinese traditions emphasized the authority of the architectural plan over other factors." According to my understanding, real "environmental influence" is the result of the interaction between the so-called "formalism" and the "natural" environment. Alternatively, politics, the patriarchal clan system, customs, etiquette, Buddhism, Taoism, and *feng shui,* which are "environmental influences" imply that the Chinese spirit is also "nature" in the broad sense.

In my observation and understanding, "artificiality" and "nature" are not static, but can be transformed into each other under certain conditions. Artificial objects can have some naturalness, while natural things can also be "made" by mixing artificial and pure natural elements. The blurring boundary between man-made and natural can produce "naturalized artificiality" and "artificialized nature" through mutually derivation and transformation. The "artificial" ideal depends on nature and beautiful examples of "nature" can be manipulated by man.

The ideal architecture we strive for should provoke a spiritual reaction in people. There are many ways for architecture to inspire: its monumentality and tectonics can harmonize solemnity and materiality. The peach gardens described by the ancients and the aspirational lifestyle depicted in John Hejduk's "paper architecture" can bring forth the "poetic idealism" . I chose the approach of using the "interaction between architecture and nature" (*jianzhu yu ziran jiaohu*) to produce my poetry. "*Jiaohu*" refers to the interaction and association of two opposite, yet and complementary things.

The goal of "Integrated Geometry and Poetic Scenery" is to create buildings that interact with "nature". The diverse examples of "nature" in contemporary reality requires corresponding "artificial" strategies to interact with it. The "nature" we encounter in practice varies among different regions, environments, and historical sites and present rich differences in "nature". Therefore, the observation, perception, research, and evaluation of the "natural" state inside and outside of a site are crucially important. How do we consider various "natural" states and what architectural strategies can we use to interact with them? In the ten years since 2010, our architectural practice has focused more on this issue, presenting different states of "nature" and examining its interplay with architecture in an attempt to create a variety of ideal living spaces for people depending on the user's unique, contemporary realities.

Characteristically, the sites where the Jixi Museum and the Gymnasium of the New Campus of Tianjin University are located are at the opposite ends of contemporary Chinese cities. Jixi is a place with "rich nature" (*fengrao ziran*) typical for a historic, thousand-year old town with both "primordial nature" surrounded by mountains and rivers, and "artificial nature". The New Campus of Tianjin University is a piece of blank slate in the new city rising from the ground without any sense of place or belonging, which can be called as "blank nature" (*kongbai ziran*). Given the "rich nature" surrounding Jixi Museum, we adopted the method of "appropriation and management" (*yinjie yu jingying*), to make the building's internal and external spaces, structures, and forms interact with the "primordial nature" of landscapes and trees and the "artificial nature" of residential housings in the town. Our aim was to allow people to experience a harmonic environment that transcends time, space and life. For the "blank nature" associated with the Gymnasium of the New Campus of Tianjin University, it was necessary to create an "artificial nature" composed of structural "settlements". The interaction between the Gymnasium's "blank nature" and the "special nature" of human movement within fills the void and creates a sense of belonging and place.

型的北京旧城大杂院，历史悠久但日趋失序、破败，存在诸多矛盾和问题，同时又有改造的局限，我视之为"历史自然"；而唐山则是因大地震后快速重建而丧失时间痕迹的新城，现代而乏味，我称之为"乏味的城市自然"。对于大院胡同28号所面对的"历史自然"，采用"分型加密、重建规制"的方式，将大杂院转变为"微缩社区"，使人们重归日常生活中的"宅园诗意"，并时时体验到老城所特有的"都市胜景"，受限于旧城的高度限制，在水平维度展开改造性的"聚落"营造；而对于唐山"第三空间"所面对的"乏味的城市自然"，则需要通过居住空间单元的错层结构形成人工台地，带来丰富的生活体验，建筑立面的悬挑亭台收纳乏味的城市景观并将其转化为日常生活中的动人画面，建筑自身也成为城市中的新景观，在高密度的城市建造环境中，向垂直维度发展，成为向高空延伸的立体城市聚落。

首钢工舍智选假日酒店和延庆园艺小镇文创中心所在的场地，是当代中国城市不同发展方式所体现的两极案例。首钢园区是工业时代的辉煌见证，是大规模钢铁生产需求下建造的特殊"城市"，随着时代落幕，这里原有的返焦返矿仓和高炉空压机站、配电室、转运站长期荒芜几近废墟，我视之为"废墟自然"；延庆2019世园会所在地原是几个风貌朴质、自然丰饶的村庄及农田，被"扫荡"式拆除后作为会场展示用地，虽然重新种植的各种展示花木是真的自然植物，但我内心依然排斥这个非原生的、刻意造作的"布景式自然"。对于首钢工舍所面对的"废墟自然"，采用"珍视保护，新旧相生"的策略，将工业遗产作为时间和记忆的载体，被保留的旧仓与叠加其上的新阁并置——"上阁下仓"，当下新的建造与"废墟自然"交互对话、相融共生；对于延庆园艺小镇文创中心所面对的"布景式自然"，则采用"内向叙事"的策略，表达一种批判性的态度，与周边"不自然"的布景式建筑和景观保持距离，"筑房拟山，自我交互"，两组单坡屋面赋予建筑强烈的方向性和识别性，并完成了象征自然物的"山"与象征人工物的"房"之间的连接和互成。

可见，"自然"并非天然存在的一种全然美好，需要人工即时即地的合理介入。对于"丰饶自然"，可作因借与经营以将诗意关联凸显；对于"空白自然"，则需要填充与自造以建构归属与场所；对于"历史自然"，重在回归起源以复现与新生；对于"乏味的城市自然"，则要通过人工的引导与框界加以激活与赋能；对于"废墟自然"，强调珍视与利用，使得新旧交互，相融共生；对于"布景式自然"，则可能需要人工的"反介入"，达成与自我的交互，这同样也是消弭"虚假自然"，缔造新的"自然"可能性的一种方式与途径。在所有那些关于现实中理想空间和诗意胜景的营造中，人工与"自然"密切互动相融的关系都是最为重要与关键的主题。

面对当代世界特别是中国生活环境与空间中人工与自然失衡、割裂的严峻现实和跨越广泛的地域、文化等条件的多样现实，通过长期的建筑思考和实践，我努力归纳出"现实理想空间范式"营造的五种要素：风水形势、叙事空间、结构场域、人作天工、胜景情境。"现实理想空间范式"的核心是"胜景几何"——以结构、空间、形式、建造等几何为基础互动衍化的建筑本体元素，营造人工与自然交互相成的空间情境，使之成为使用者的理想生活空间。

"胜景几何"的核心是"与自然交互的建筑"，而与"自然"交互的建筑，并不仅是物化的建筑本身，更是通过建筑作为人与自然交互的媒介，使人这一核心"主体"突破物质时空的限制，获得"身临其境"的诗意情境，实现人们对现实和未来理想生活空间的向往和可能性。

与"自然"交互的建筑，作为人与自然交互的中介，也会获得一种令人愉悦的自然感——一种"自然性"。我认为这种"自然性"，是对于身处严峻、多样之现实中的当代人获得理想之生存和生活空间至关重要的建筑要素和原则之一。

两千多年前，古罗马建筑师维特鲁威（Vitruvius，公元前1世纪）在《建筑十书》中提出了经典的建筑三原则：坚固、适用、愉悦（Solidity、Utility、Delight）。以"胜景几何"为基础，如果可以尝试对于传统建筑学的范畴进行当代的修正扩展，那么我愿意将"自然"纳入，使之成为新的"建筑四原则"：坚固、适用、自然、愉悦（Solidity、Utility、Naturality、Delight）。自然，被强化为与空间、结构、形式、建造等同等重要的建筑学本体要素，如此或将可以提出一种面向现实和未来的当代建筑学，助力我们进入人的生活空间营造的新境界。

The sites of No. 28 Beijing Dayuan Hutong and the "Third Space" in Tangshan are two polar opposite examples of contemporary Chinese urban residential spaces. The Dayuan Hutong is a typical compound in Beijing's old city. It has a long history, but is increasingly disordered and dilapidated. With its many contradictions and problems, it has, the same time, limitations for rehabilitation. I regard the Dayuan Hutong as "historical nature". Tangshan, rebuilt as a new city after the 1976 earthquake, has lost its sense of time. Tangshan is modern, yet tedious. I deem it "boring urban nature". Regarding the "historical nature" faced by No. 28 Dayuan Hutong, the method of "increasing density according to fractal encryption and reconstructing regulation" (*fenxing jiami, chongjian guizhi*) was adopted to transform the large compound occupied by many households (*dazayuan*) into a "mini-community" so people could experience not only the the daily pleasures of a home-garden (*zhaiyuan*), but also the unique "urban poetic scenery" of the old city from time to time. Limited vertically by the height restrictions of the old city, the renovation designed horizontally. To improve on the "boring urban nature" of the "Third Space" in Tangshan, we designed artificial terraces for each living space unit, which invited a more environmentally inclusive living experience. The cantilevered pavilions on the building's facade accommodate the static urban landscape and transforms it into a a more vibrant reflection of daily life. The building itself has also become a new landscape in the city. The project emerges vertically from a dense urban, horizontal landscape resulting in a striking three-dimensional community extending to the sky.

The "Silo Pavilion", Holiday Inn Express Beijing Shougang and Mountain Garden, Cultural Center of Horticulture Village in Yanqing reflect different developmental modes in contemporary Chinese cities. The Shougang industrial heritage park is a unique industrial era "city" built during a period of demanding for large-scale steel production. With the end of that era, the original silo, blast furnace, air compressor station, distribution room and transfer station were long abandoned. I consider them as "ruined nature". The location of the International Horticultural Exhibition 2019 originally contained a collection of villages amidst rich farmland, which were demolished and transformed into the exhibition site. Although replanted with ornamental flowers and trees that appeared native, I still could not understand its intention and rejected this contrived "set nature". For the "ruined nature" the "Silo Pavilion", we adopted a strategy of "preserving the old to coexist with the new". We juxtaposed new and old by superimposing a new pavilion over the existing bunker. The "Silo Pavilion's" industrial heritage is preserved as the new elements interact and coexist with this "ruined nature". For the "set nature" of the Mountain Garden, we adopted an "introverted narrative" (*neixiang xushi*) strategy to cast a critical eye towards and to distance it from the surrounding contrived buildings and landscapes. The Mountain Garden's two sets of single-slope roofs give the building a strong directionality and identity. In between, "houses" symbolizing the man-made connect the "mountains" that symbolize nature and I called it "building a house and simulating a mountain, and realizing self-interaction (*zhufang nishan, ziwo jiaohu*)".

"Nature" is not necessarily inherently beautiful, but can benefit from reasonable human intervention in some circumstances. For "rich nature", it is possible to adopt the method of "appropriation and management" to highlight its poetic connection. For "blank nature", it is necessary to fill, create and build the sense of belonging and place. For "historical nature", the emphasis is its origins, its rebirth and its regeneration. For "boring urban nature", it needs to be stimulated and empowered through artificial guidance and frame. For the "ruined nature", emphasis should be placed on reverence and use, so that the old and the new can interact and coexist. For "set nature", it is possible to achieve self-interaction through artificial "anti-intervention", which is also a way and approach to prevent false nature and create new natural possibilities. In the creation of ideal space and poetic scenery in reality, the close interaction between artificiality and "nature" is the key theme.

Faced with the harsh reality of the imbalance and fragmentation between man and nature in the contemporary world, especially in China, I strive to summarize, through long-term architectural thinking and practice, the five elements comprising the "paradigm for realistic ideal space": *feng shui* situation, the narrative space, the structural field, the man-made and nature-work, the situation and the poetic scenery. The core idea of "paradigm for realistic ideal space" is "Integrated Geometry and Poetic Scenery" that integrates the ontological architectural elements that evolved from geometry, including structure, space, form, and construction, to create spatial interactions between artificiality and nature resulting in ideal living space.

The core of "Integrated Geometry and Poetic Scenery" is "architecture that interacts with nature". When architecture serves as a means for human interaction with nature, human beings can break through the limits of materials, time and

建筑中自然、空间、结构、形式、建造等要素的互动转化，实质即人工与自然的交互，或可更加适应于当代现实中人的生存和生活理想，这些恒久不变的要素是建筑学的本体价值和意义所在，对于它们的持续思考和建筑表达，将使我们更好地面对严峻、复杂、多样的人类现实和未来。

我的思考起点是中国和东方的传统城市、建筑乃至文化、哲学，而我希望思想和行动的方向可以超越地域、文化和时代，指向过去、现在和未来共同的人性中人类可以共享的生活理想。

中国文化中的"自然"还有另外一种含义：一种自然而然、无为而自成、任运的状态。"人法地，地法天，天法道，道法自然"，老子一气贯通，揭示宇宙天地间万事万物均效法或遵循"道"的"自然而然"之规律（其实这一思想与"自然"的拉丁文含义"天地万物之道"亦有相通之处），"道"以自己为法则。"人为道能自然者，故道可得而通⋯⋯是故凡人为道，当以自然而成其名。"在这里，"自然"更多的是指思考的境界与实践的方法，使得关于建筑本体之"自然"的内涵更加深刻而全面。希望在不远的将来，我的建筑思考和实践可以进一步抵达更为深刻而全面的"自然"之境界。

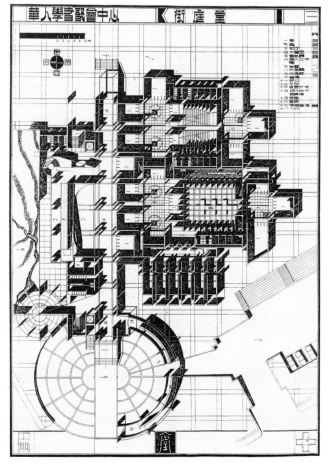

华人学者聚会中心（毕业设计，1991）
The Chinese Scholars Center (Schematic proposal for graduation design, 1991)

space, to attain the "immersive" poetic environment, and aspire to create the ideal living space.

Architecture that interacts with "nature", serving as an intermediary between man and nature, also acquires a pleasant sense of nature—a kind of "naturalness". In my opinion, this "naturalness" is one of the architectural elements and principles essential for human beings to achieve harmony with their environment.

More than 2,000 years ago, the ancient Roman architect Vitruvius proposed three classic principles of architecture in his *The Ten Books on Architecture* : solidity, utility, and delight. From the concepts I developed as "Integrated Geometry and Poetic Scenery," I propose a contemporary amendment to Vitruvius's classical categorization of architecture by adding "nature" resulting in "four principles of architecture": solidity, utility, naturality and delight. Nature is underlined as important as ontological architectural elements such as space, structure, form, and construction, and guides contemporary architecture towards the realities of the future and helps us enter a new realm of living.

The interaction of nature, space, structure, form, and construction is essentially the interaction between the artificial and the natural. Transforming our environments allows us to adapt to contemporary realities. The ontological value and significance of architecture lie in these foundations. The continuous thought and architectural expression of such interactions will enable us to better face the grim, complex, and diverse future of humanity.

Although the starting point of my thought is Chinese and Eastern traditional cities, architecture, culture, and philosophy, I hope that the progression of thought and actions can transcend regions, cultures, and time across humanity.

Another meaning of "nature" in Chinese culture is the state of being natural, without harming nature, but still achieving self-sufficiency. "Man follows the earth. Earth follows heaven. Heaven follows the Tao. The Tao follows what is natural." Lao Zi coherently revealed the "natural" law of Tao. Everything in the universe, between heaven and earth, reflects and follows the Tao. (The Tao has something in common with the meaning of "nature" in Latin—"the way of heaven and earth.")If human beings seek the Tao naturally, the Tao is accessible. Therefore, the Tao should be practiced naturally. Here, "natural" refers more to the realm of thinking and the method of practice. This understanding of "nature" makes architectural ontology more profound and comprehensive. I hope in the near future, my architectural and practice can further reach a much more profound and comprehensive state of "nature".

2001

兴涛接待展示中心
XINGTAO RECEPTION & EXHIBITION CENTER

兴涛接待展示中心设计始于 2001 年 4 月，2001 年 9 月建成投入使用，位于北京大兴区一个开发中的商品住宅小区的入口处，包含了接待、展示、洽谈、住宅样板间、小区大门及警卫室等功能空间。这个小建筑试图将它特有的商业特征与中国传统园林的空间体验和东方意味融合在一起，用一种有趣的、传统的方式来实现商业的、现代的功能，并使用当地的、现时／现代的、可操作的技术满足业主的现实需求和低造价下的快速建造要求。

设计将整个建筑分为两组体量，分置在狭长的用地南北两端，南侧是两层体量的接待展示空间，北侧是单层体量的住宅样板间单元和象征性的小区大门及警卫室，中间隔以水池并以一条长长的线型墙／廊相连。由于展示、接待、参观、洽谈和销售的流程而产生了一种动态并循环的流线，并通过这一流线来实现建筑的使用过程，这一流线是由墙体要素的延伸变化来引导的。墙对空间体验的动态引导性是中国传统建筑特别是古典园林的重要特点之一，在中国古典园林中，由于连续的墙体所特有的导向性，使身处其中的人不由自主地产生一探究竟的欲望，由此在人的运动中发生丰富的空间体验，使中国园林成了真正的"四维建筑"。在这里，一片白墙由建筑的入口开始，不断地在水平和垂直方向延伸运动，忽而为垂直的墙，忽而是水平的板，或升或降，或高或低，如此形成了建筑的骨架和内外空间；在这个建筑和空间骨架中再插入透明的玻璃体和玻璃廊、灰砖的样板间单元和警卫室以及一片黑色的浅水池。

室内的家具和展板，室外的景观水面和庭园，乃至小区的建筑风貌，都与建筑的功能流线和使用者的空间体验有着密切的关系。建筑语言和材料的使用力图体现出现代感和某种抽象性，以凸显人在内外空间中的运动、视觉和身体体验。

摄影 / Photographer:
张广源 / Zhang Guangyuan

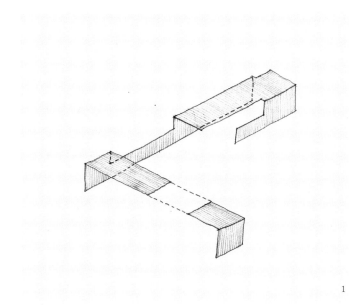

1

2

The design of Xingtao Reception and Exhibition Center began in April 2001. It was put into use in September 2001. Located at the entrance to a residential community in the Daxing District of Beijing, it provides space for reception, exhibition, sales offices, a residential model unit, a security guard office, and an entrance gate for the community. This small building tries to integrate its commercial space and the Eastern influence of traditional Chinese gardens, so that the modern, commercial function of the space can be accommodated in an interesting and traditional manner. Use of locally available building materials and state-of-the-art technology met the practical needs of the owners who required a low-cost, quickly constructed building.

The building is divided into two wings located at the northern and southern ends of a long, narrow site. The southern wing houses the two-story reception and exhibition space. The single-story northern wing contains the model unit, the community entrance, and the security office. The two areas are separated by a reflecting pool, but connected by a long pedestrian corridor set against an equally long wall. With its heights changing and lengths extending, the wall guides visitors through the reception area, along the walkway next to the reflecting pool, and into the sales spaces. Using walls to subtly guide visitors through a space is a characteristic of traditional Chinese architecture. In particular, classically designed gardens use long but unimposing walls along walkways to gently encourage visitors through a predetermined path. As visitors move through these gardens, they are immersed in the space resulting in "four-dimensional architecture". This is just the case in Xingtao Center, where a white wall beginning at the entrance extends in the horizontal and vertical directions, sometimes becoming vertical walls and sometimes horizontal slabs, rising up and down, high and low, and eventually forming the building's internal and external skeleton. Along this foundational skeleton, we placed glass structures, a long corridor, the model unit, the security area, and the shimmering reflecting pool.

The furniture, the display boards, the pool, the rear garden, and even the buildings' layout of the community are related to the building's functions, the movement of users, and users' spatial experience. Through the language and materials of architecture, we tried to embody a kind of modern and abstract senses, so as to highlight the movement, visual experience, and spatial orientation of visitors amidst internal and external spaces.

3

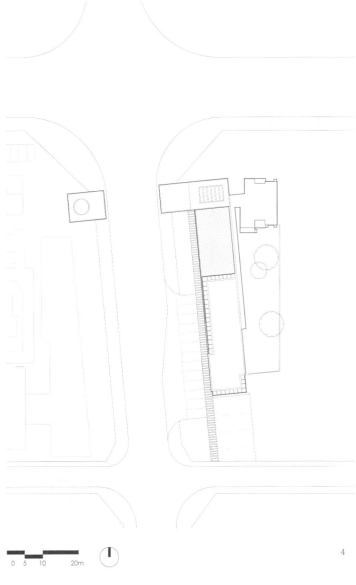

0 5 10   20m

4

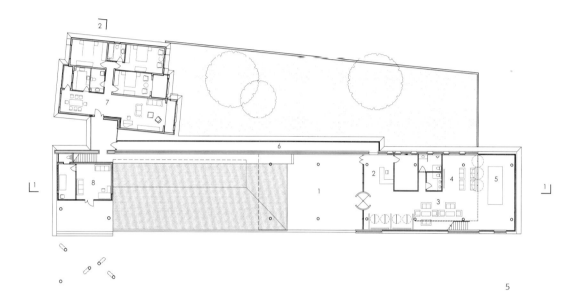

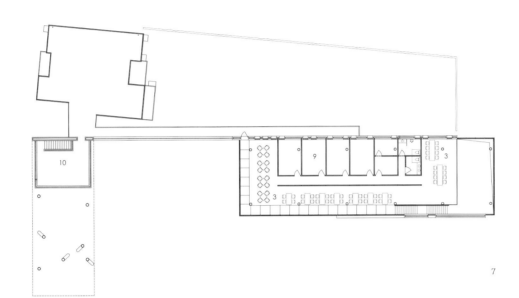

7

1 入口 / Entrance
2 接待区 / Reception
3 洽谈签约区 / Signing area
4 放映区 / Video show
5 模型展示区 / Model area
6 玻璃廊 / Glass corridor
7 样板间 / Simple unit
8 警卫室 / Guard
9 办公室 / Office
10 屋顶平台 / Roof terrace

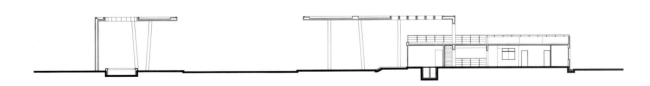

8

9 北立面 / The north elevation

10 由水池望向入口 / View from the pool to the entrance

11 由水池望向小区大门 / View from the pool to the gate of the residential community

0 2   5   10m

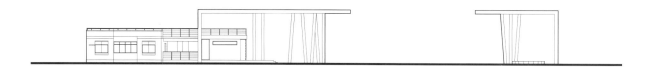

9

10

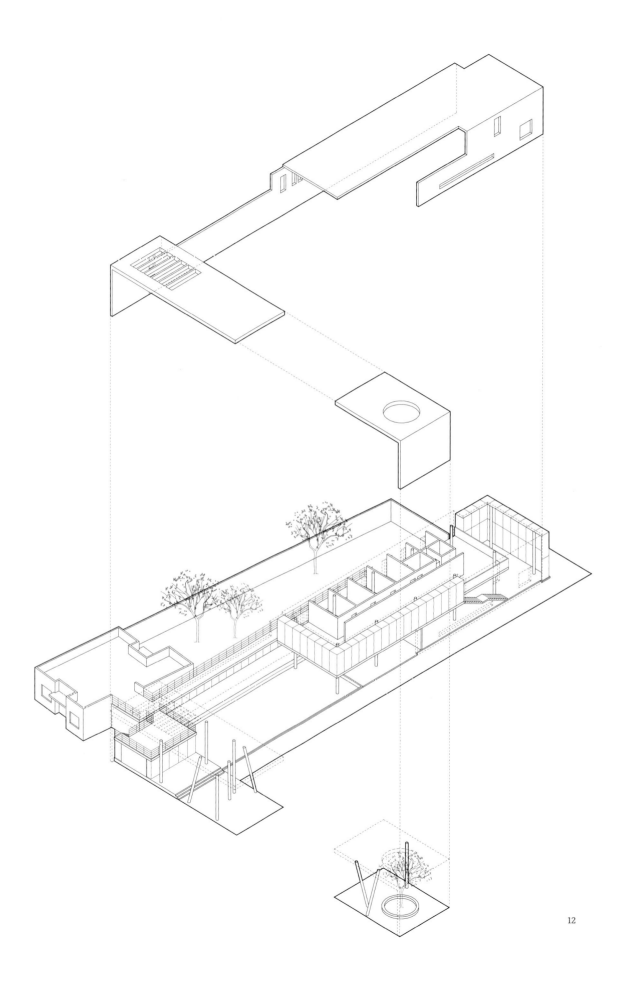

12

13

2007

北京复兴路乙 59-1 号改造
RENOVATION OF
NO. B-59-1, FUXING ROAD
IN BEIJING

北京复兴路乙 59-1 号改造项目设计始于 2004 年 10 月，2007 年 5 月建成投入使用，位于北京长安街西延长线复兴路北侧，原是一幢 20 世纪 90 年代初期设计建造的、9 层钢筋混凝土框架结构的办公和公寓建筑，并在东侧与一栋 9 层住宅楼相连。业主希望在保持原有建筑高度、结构、设施基本不变的基础上，对功能、空间和外观进行整理改善，将其改造为集餐饮、办公、展廊为一体的小型城市复合体。

改造后的建筑体形基于原来的方形体量，并结合周边环境和日照关系进行局部的切削和增长；改造后建筑外幕墙框架网格的生成基于原有建筑较无规律的立体结构体系（自下而上的不规则层高分布和从左至右、由前到后变化多样的柱距），并用来作为立面及内部空间的控制系统，貌似不规则的幕墙网格实际是原有建筑结构框架的体现和强调，既符合改造加建的结构逻辑，天然地反映着原建筑的基本状况，又形成有自身独立特征的结构和形象语言。将生成的外部幕墙网格立体化和空间化，幕墙由原框架结构分别向外做不同尺度的悬挑，在不同朝向形成不同进深和特征的内部空间，以配合不同的使用和景观要求。西侧利用原室外消防疏散楼梯，将其扩展改造成为一个拥有十层高度的立体画廊，不同高度、位置、形态和景观的展厅和平台被多样的楼梯、踏步和台阶联系起来，由下至上，一直延伸到局部加建的顶层及屋顶庭院，画廊内部实体部分和虚空部分的多样形态，完全基于被空间化的幕墙网格，可被视为一个垂直方向游赏的小型园林，让人领略不同高度的内外景物，体验曲折多变、空间开阖的繁复过程。

对应内部功能和空间的不同，选用了四种不同透明度的白色彩釉玻璃，作为覆盖在外部网格上的幕墙材料，既控制着光线在建筑内外的投射和透射，以塑造不同氛围的室内空间和包容不同通透表面的室外形体，也左右着人的视线在建筑内外的驻留和延伸，以观赏展品、空间和不同清晰度的城市景物，并使街道上的人由不同透明度的表面留意到建筑内部的不同景象。不均匀的透明度加上全隐框的白色彩釉玻璃幕墙，使得建筑呈现出深邃、平静而丰富的气质。

摄影 / Photographer:

张广源 / Zhang Guangyuan

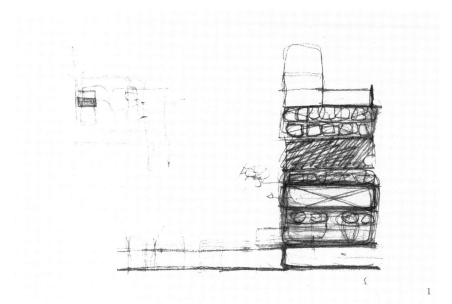

1

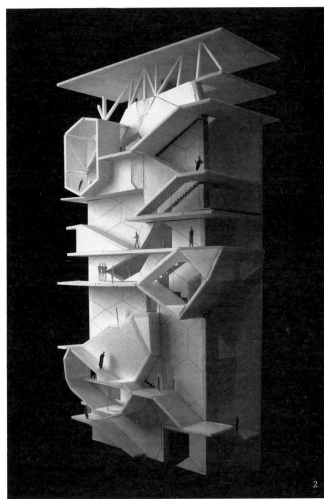

2

The renovation project of No. B-59-1 of Fuxing Road was designed in October 2004 and finished in May 2007. It is situated on the north side of Fuxing Road, on the western extension of Chang' an Avenue in Beijing. It was originally a 9-story office and apartment building built in the 1990s with reinforced concrete. Connected to this main structure is a 9-story residential building to the east. Without altering the building's height, structure, or facilities, the owner tasked us to transform it into an urban complex complete with a restaurant, offices, and a gallery.

The renovated building is based on the existing footprint and structure, but partially modified to allow neighboring buildings access to sunlight. After the renovation, I based the design of the exterior facade on the original building's irregular framework, which included different floor heights and varied column spacing in all directions. The existing structure continues to serve as the guide for the new facade and internal spaces. The seemingly haphazard gridded, glass curtain facade is actually hearkens the original building's structural frame, while still satisfies the owner's renovation requirements. They not only comply with the former structural logic, but also form individual structural and visual languages. The newly constructed external curtain facade is cantilevered with different dimensions than the original frame structure. The grids create interior three-dimensional spaces offering different depths and characteristics to meet varying functions and spaces. On the west side, the former outdoor emergency stair is expanded to a stereoscopic gallery with ten-floor height. The exhibition halls and platforms of different heights, shapes, and views extend from the ground floor to the partially modified top floor and roof courtyard. The positive and negative space of the gallery is outlined by the irregular shapes formed by the exterior glass curtain. The gallery can be regarded as a small garden with vertical direction, allowing people to appreciate the internal and external scenery of different heights, and to experience the complicated process of twists and turns and space opening and closing.

To further delineate uses of interior spaces, for the glass facade, I selected four types of glazed, white glass with different opacities. The different glazings allow for varying casts of light and shadows to change the interior atmosphere. The exterior glass shell projects light and shadow into the interior galleries to allow pedestrians to pay attention to the different scenes playing out within the building. The fluctuating transparency of the glass curtain frames the building and communicates a tranquil, yet rich temperament emanating from within its walls.

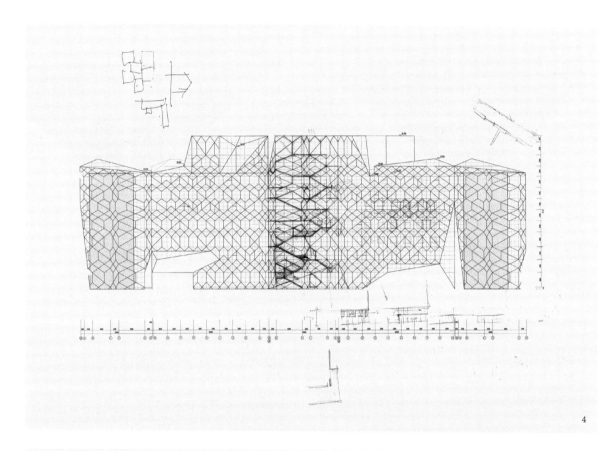

4

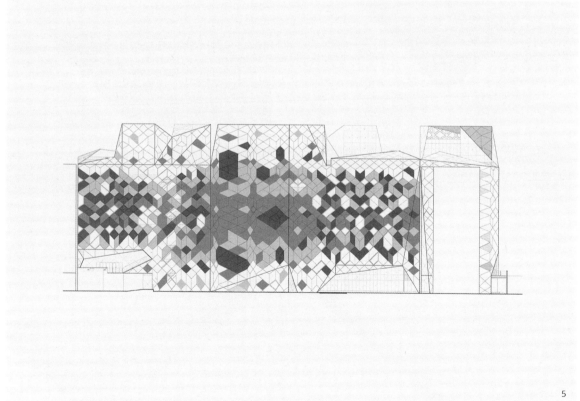

5

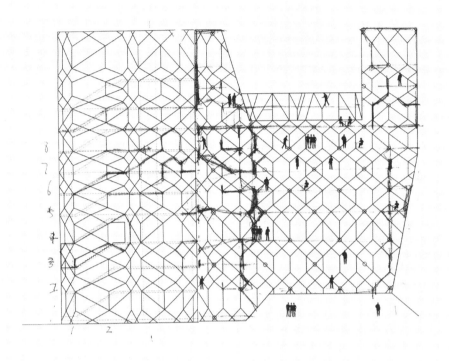

6

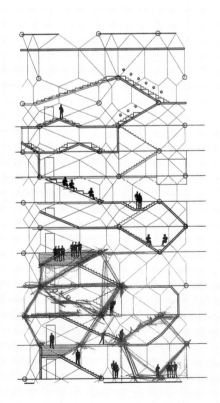

7

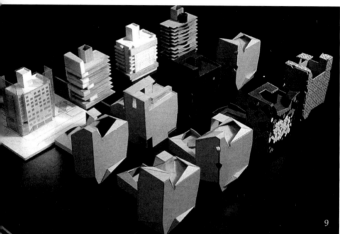

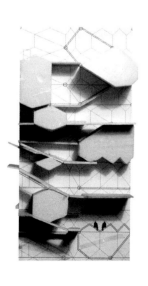

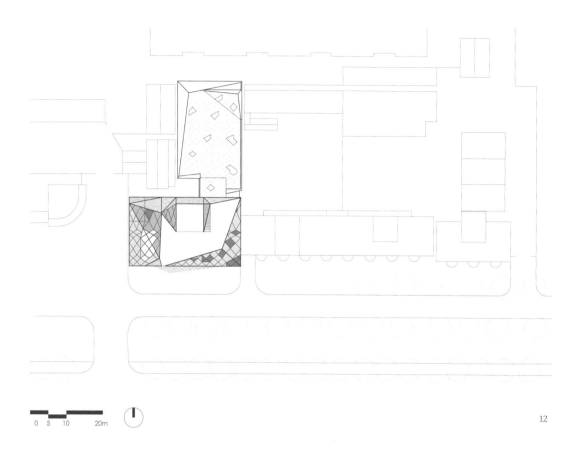

0 5 10 20m

12

13

15 西立面局部 / Detail of the west facade
16 一层平面 / The 1st floor plan
17 剖面 1-1/ Section 1-1

0 1 2   5m

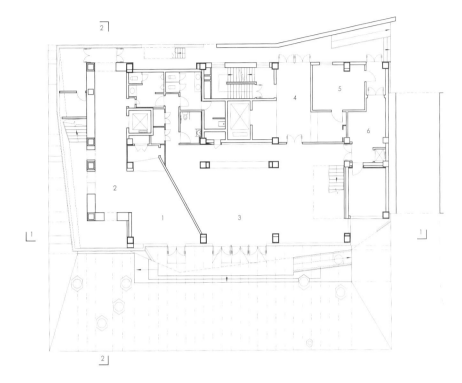

16

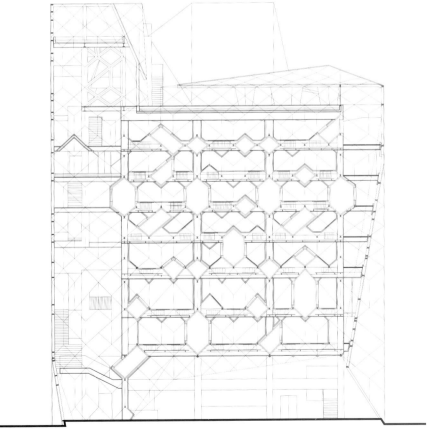

1 画廊门厅 / Lobby of gallery
2 立体画廊 / Stereoscopic gallery
3 咖啡厅 / Cafe
4 办公门厅 / Lobby of office
5 消防控制室 / Fire control room
6 预留厨房 / Reserved for kitchen
7 办公区 / Office
8 休息区 / Break area

17

18 标准层平面 / The standard floor plan
19 剖面 2-2/ Section 2-2
20 立体画廊人视 / View of the stereoscopic gallery

0 1 2   5m

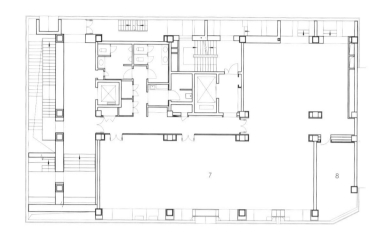

18

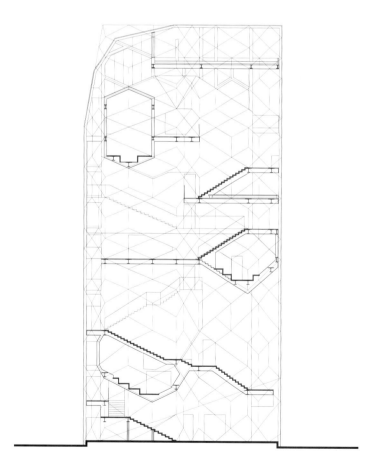

19

21

硬木条地板 / Hard wood slat floor
木龙骨 / Wood keel
240*240 砖墩，纵横间距 1000/ Brick block@1000
40 厚 C20 细石混凝土垫层 / 40 concrete
50 厚挤塑聚苯板 / 50 XPS
3+3SBS 改性沥青防水卷材 / 3+3 SBS
20 厚 1:3 砂浆找平层 / 20 cement mortar
最薄 30 厚水泥粉煤灰页岩陶粒找坡层 / 30
cement ashcement ceramsite slope
原钢筋混凝土屋面板 / Concrete roof

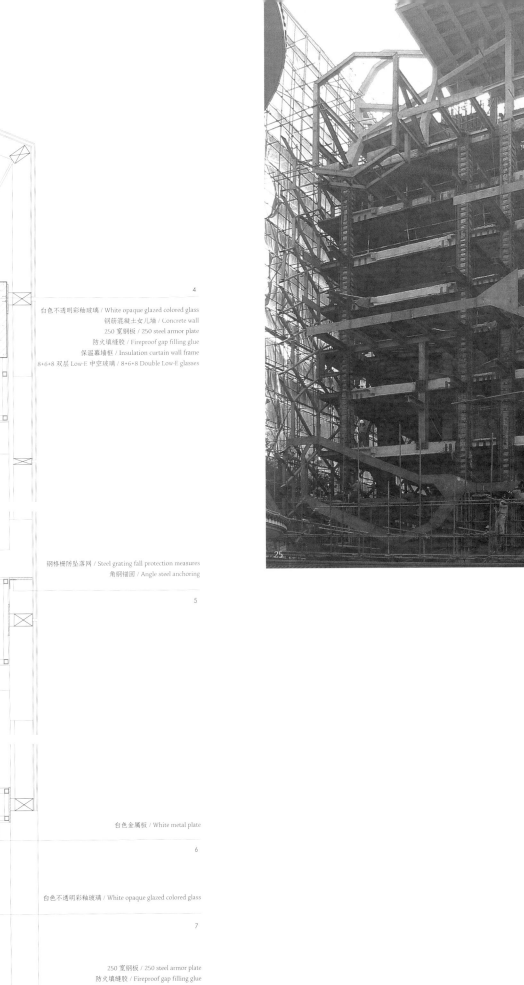

4

白色不透明彩釉玻璃 / White opaque glazed colored glass
钢筋混凝土女儿墙 / Concrete wall
250 宽钢板 / 250 steel armor plate
防火填缝胶 / Fireproof gap filling glue
保温幕墙框 / Insulation curtain wall frame
8+6+8 双层 Low-E 中空玻璃 / 8+6+8 Double Low-E glasses

35-70 厚灰色水泥白流平涂料 / 35-70 gray cement selfleveling
封闭剂一道 / Blocking agent
10 厚自流平厚质涂料涂层 / Selfleveling paint coating
25 厚水泥砂浆抹面压实赶光 / 25 cement mortar
原钢筋混凝土楼板 / Concrete roof

钢格栅防坠落网 / Steel grating fall protection measures
角钢锚固 / Angle steel anchoring

5

石膏板吊顶 / Gypsum board
轻钢主龙骨 / Light steel main keel
轻钢次龙骨 / Light steel sub keel
纸面石膏板 / Gypsum plaster board
封底漆一道 / Seal primer
合成树脂乳液入料面层 / Synthetic resin emulsion paint

白色金属板 / White metal plate

6

白色不透明彩釉玻璃 / White opaque glazed colored glass

7

250 宽钢板 / 250 steel armor plate
防火填缝胶 / Fireproof gap filling glue
保温幕墙框 / Insulation curtain wall frame
8+6+8 双层中空玻璃 / 8+6+8 double glasses

8

24

2008

威尼斯纸砖房
PAPER-BRICK HOUSE
IN VENICE

威尼斯纸砖房设计始于2008年5月,2008年9月建成投入使用,2009年12月撤展拆除,位于威尼斯军械库(Arsenale)处女花园,是应第11届威尼斯国际建筑双年展中国国家馆策展人邀请设计建造的参展作品。本届双年展中国馆的主题是"普通建筑",当代中国大量、快速的城市建设中,量大面广的普通建筑缺乏应有的设计建造质量和建筑师的真正关注,并渐渐失去与自然之间的合理关系。纸砖房作为中国馆参展并实地建造的五栋"普通建筑"之一,以建筑师所在的大型国有设计院日常装输出图纸的纸箱作为"纸砖"砌筑"纸墙",以日常打印设计图纸剩余的打印机用纸轴(纸管)作为"纸梁"搭建门窗过梁和楼板、屋顶,从而用纸材料建造起一所可供坐卧起居、游戏会客、阅读静思等日常活动的房子。

一方面,纸砖房是对汶川大地震中钢筋混凝土建筑及其质量问题使其成为埋葬活人之坟墓的悲剧的直接反应——为什么建筑不能轻一些从而安全一些呢? 这也许意味着应对自然的不同方式和建筑另外的发展方向:以柔和的方式应对自然的轻型建筑,而非以抵抗的方式应对自然的重型建筑。另一方面,纸砖房使用了令人目眩的大量图纸箱和打印纸轴,暗示着当下中国生产式输出的建筑设计状态,提示中国建筑师和在中国工作的外国建筑师在应对大量、高速的城市建设的同时,必须面对的来自质量控制的挑战以及应对策略;同时,使用某种"预制"和重复的"标准化"元素作为建筑单体及材料构件,可以成为较为理想的建筑质量控制手段,这也是中国悠久的建筑传统之一。纸砖房中内向性的庭院空间是建筑的核心,这也来自中国的传统;同时,建筑也加强了与街道及相邻建筑的关联,并提供外部行人停留休息使用的公共空间和设施,使其具有了更多的公共性。

纸砖房的建造方式非常直接:包括"纸砖"的强化/防水构造和砌法、"纸管梁"的联结/防水构造和与砌体的联结等。由于现场不能做地下基础,所以设计了浮置在网眼编织袋(内装级配砂石)上的纸管"筏板"做法,这也与威尼斯在海中淤泥层上建造漂浮城市的传统暗合,是一种典型的减震构造。

摄影 / Photographers:
孙鹏、李宁、李兴钢、付邦保 / Sun Peng, Li Ning, Li Xinggang, Fu Bangbao

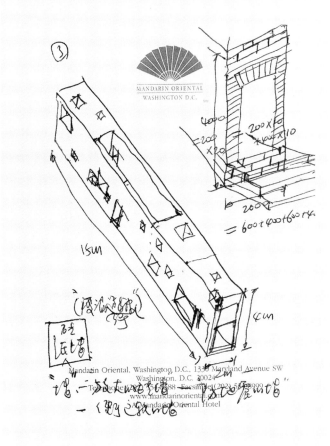

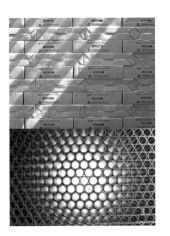

The Paper-brick House in Venice was designed in May 2008, completed, put into use in September 2008, and removed in December 2009. Located in the Virgin Garden of the "Arsenale" in Venice, it was commissioned by the curator of China Pavilion for the 11th Venice International Architecture Biennale. The theme of China Pavilion was "ordinary architecture". During the massive and rapid urban construction period in contemporary China, a large number of ordinary buildings lacked proper design, construction quality, or attention from architects. Gradually, these buildings lost any relationship with nature. The Paper-brick House was one of the five "ordinary buildings" built and exhibited on site for the China Pavilion. Using paper materials, we built this house for living, entertaining, socializing, reading, and reflection. In fact, the walls were built with "paper bricks" made of cartons that were daily used to contain building design drawings at the China Architecture Design & Research Group, a large state-owned design institute for which the architect works.

The beams, floor slabs, and roof were built with "paper beams" made of disposed paper tubes from printing paper rolls used to print design drawings. The Paper-brick House was a direct response to the tragedy of the Wenchuan Earthquakes, in which reinforced concrete buildings that were not up to code entombed its victims. Instead of resisting the strength of nature with mass, an alternative method of coexisting with nature is to adapt to it by creating lightweight buildings. Additionally, the Paper-brick House uses a great deal of drawing cartons and printing paper reels that reflects the status of the contemporary construction processes in China. For Chinese architects and their foreign peers, responding to the high speed and large amount of constructions, we must overcome the challenges in quality control. Fortunately, China has a long architectural tradition of prefabrication and use of standardized construction elements, which serves as a means of quality control. At the heart of Paper-brick House was an enclosed courtyard derived from Chinese tradition. At the same time, the house maintained a strong connection with the street and adjacent buildings by providing public space for pedestrians to stop and rest.

Construction of the Paper-brick House was straightforward. It was erected using reinforcement and the masonry to bind the paper carton bricks and paper tube beams, all of which were waterproof. Because we were not allowed to dig a foundation on the site, we used the paper tubes to serve as a "raft" that floated on graded mesh bags filled with gravel. This earthquake absorption technique is in line with the building traditions of the Venetians who built their city on marshlands.

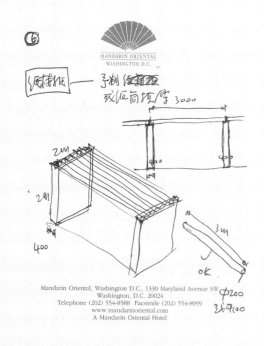

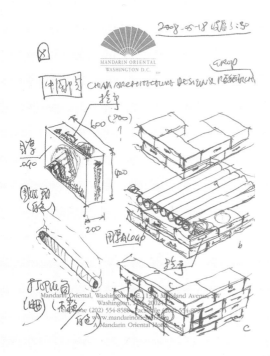

5

6

7

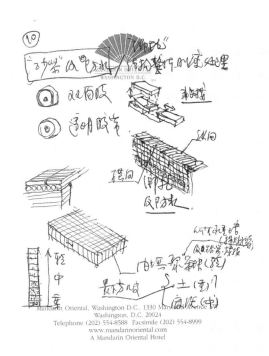

8

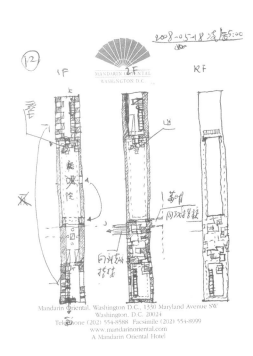

9

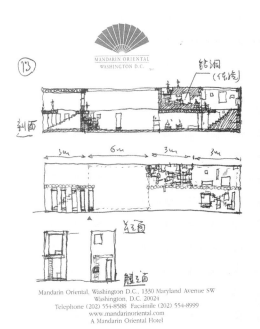

10

11 总平面图 / Site plan
12 威尼斯军械库处女花园 / View of the site
13 军械库处女花园中的纸砖房 / View of the Paper-brick House in the site

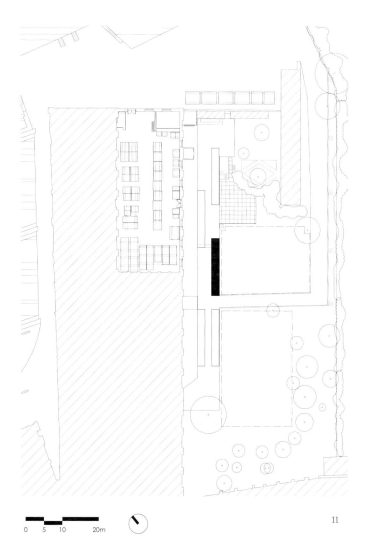

0   5   10   20m

11

12

11 总平面图 / Site plan
12 威尼斯军械库处女花园 / View of the site
13 军械库处女花园中的纸砖房 / View of the Paper-brick House in the site

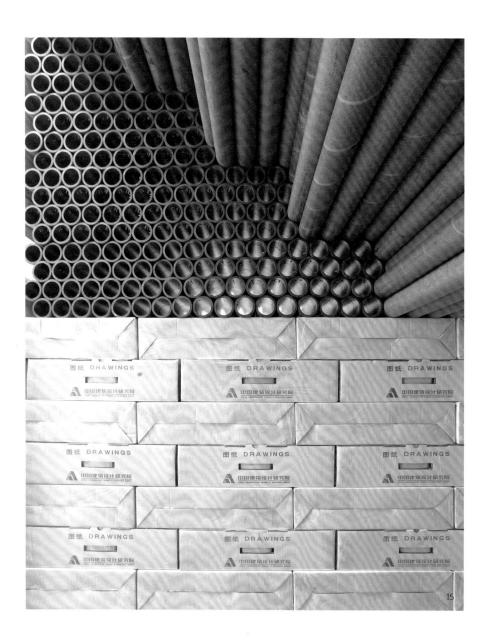

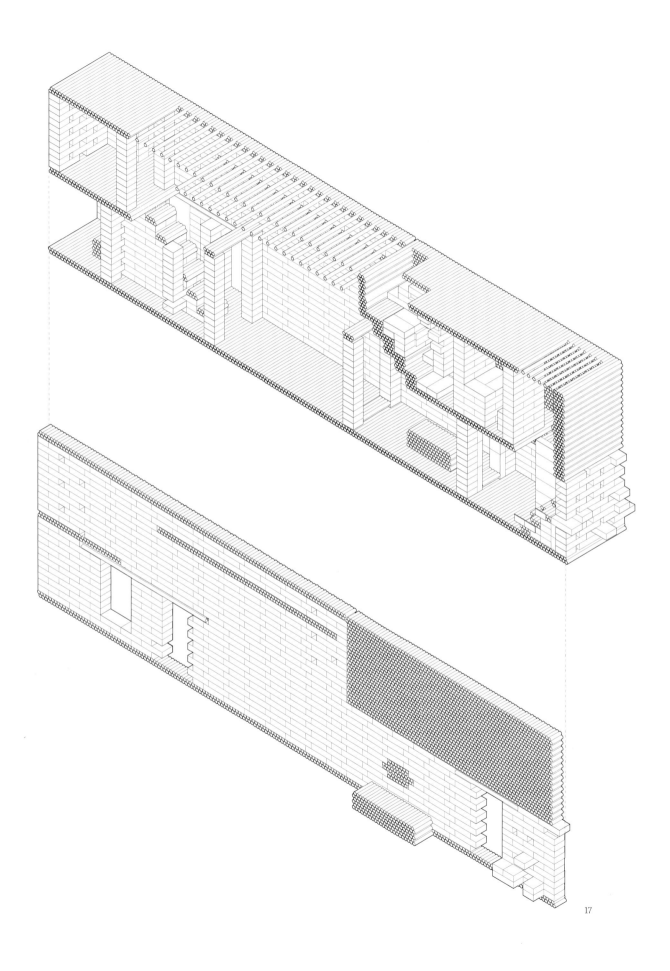

17

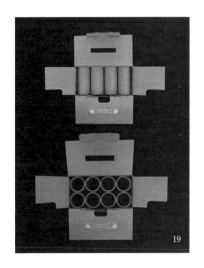

2009

李兴钢工作室室内
INTERIOR DESIGN OF ATELIER LI XINGGANG

　　李兴钢建筑工作室设计始于 2008 年 7 月, 2009 年 2 月完成投入使用, 2019 年 3 月工作室搬离并迁入新办公楼而被拆除。原工作室位于北京中国建筑设计研究院院内一栋 20 世纪 80 年代建造的小型办公楼的一个标准层。在造价限制下, 使一个原本庸常的结构和空间, 在满足使用需求的基础上, 变得有趣甚至精彩起来, 是设计的重点。

　　从功能的角度, 可以使用到电梯厅、吸烟处、大门、展廊、餐吧、设计工坊、模型制作间、材料样品角、会议 / 图书室、打印走廊、储藏间、办公室、卫生间、楼梯, 等等。从材料的角度, 可以观察到压花 / 普通钢板、欧松板 / 中密度木板、PVC 膜、自流平水泥、镜面、石膏板、软木、硬纸管、图纸箱, 等等。从构造做法的角度, 可以"阅读"到刚硬的钢板被做成柔和一体的墙地顶曲面交接, 柔软如布的膜材成为硬挺的发光顶棚, 纸管 / 纸箱"砌成"的组合墙和坐具, 墙面、大门、家具中钢和木的混合 / 组合使用, 大门 / 旋转镜门的转轴和限位构造, 石膏板吊顶的嵌入式照明, 以前旧家具的重新利用, 门 / 墙和家具 / 墙体的一体化设计, 等等。从空间游赏的角度, 可以体验或想象似有还无的水面、逆光镂空的亭榭、步移景异的游廊、透露消息的漏窗花墙、看似无用却有用的旋转镜门、狭窄却被放大的吧台、对称却很流动的空间、可以远眺城市风景的楼梯和露台, 等等; 特别是两扇旋转镜门, 不仅使光线弯折, 更使空间延伸或者旋转, 使虚拟与现实空间共存, 几乎同等地作用于人的视线、心理和意识。从精神的角度, 可以感受到现实与虚幻、实用与无用、物质与意识、理性与感性、秩序与流动……

　　上述的种种对建筑师自己工作场所的营造、解读和描述, 也大约是建筑师日常工作——思考、言说与实践中的常态: 在物质与精神、匠作与哲思、现实与理想的对立统一中煎熬而前行。这个不大的空间见证和记载了工作室的建筑师们十年间的成长记忆, 丰富而厚重, 时光和生活可以赋予空间一种特殊的"自然"状态, 需要被珍视, 并且可以去营造。

摄影 / Photographer:
张广源、李兴钢、张玉婷 / Zhang Guangyuan, Li Xinggang, Zhang Yuting

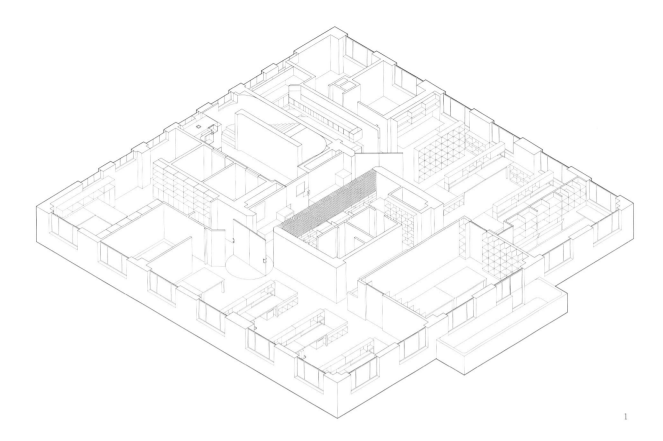

1

The Interior space of Atelier Li Xinggang was designed in July 2008 and put into use in February 2009. In March 2019, the studio was dismantled and relocated to a new office building. Located in a standard floorplan of a small office building built in the 1980s at the China Architectural Design & Research Institute, Beijing, the goal of the interior design was to convert the originally ordinary space into a more interesting, vibrant, and functional workspace, within cost constraints.

The atelier plan consists of an elevator bank, smoking pavilion, the gate, a gallery for exhibitions, bar, design workshops, a model making room, a material corner, a meeting room, a print room, a storage, offices, lavatories, and a staircase. The visible, diversity of materials used include embossed and plain steel plates, orientated strand boards (OSB), medium-density fibre boards (MDF), PVC membrane, self-leveling grout, mirrors, plasterboard, cork, hard paper tubes, and drawings cartons. From the tectonic point of view, we used curvature joints made from rigid steel plates that gently integrate the walls with floors and ceilings; seemingly rigid lighted ceilings actually made from membranes as soft as cloth; composite walls and seats made from paper tubes and cartons; composite steel and wood walls; recessed lighting in the plaster ceiling. The space also provides a setting where one can imagine a pool, a backlit cutout pavilion, a veranda where one can step into the scenery, windows and walls through which information can be exchanged, and staircase and terrace where one can overlook the city-scape. Significantly, the reflective rotating doors not only deflect light, but also extend or rotate the space to an extent that the virtual and physical space coexist with each other and appeal to human sight, mind, and consciousness equally. Spiritually, one can feel reality and illusion, utility and futility, substance and consciousness, sense and sensibility, along with order and flow.

The architect's construction, interpretation, and description of his own workplace as above may reflect how he thinks, speaks, and performs daily work: struggling and moving forward to reconcile substance and spirit, a craftsman's skill and philosophical thought, realism and idealism. This small space has witnessed and recorded the rich and bountiful growth imagination of this studio's architects for ten years. The passage of time and creative activity has contributed a special "natural" state to the space, that should be treasured and built upon.

0 1 2　　5m

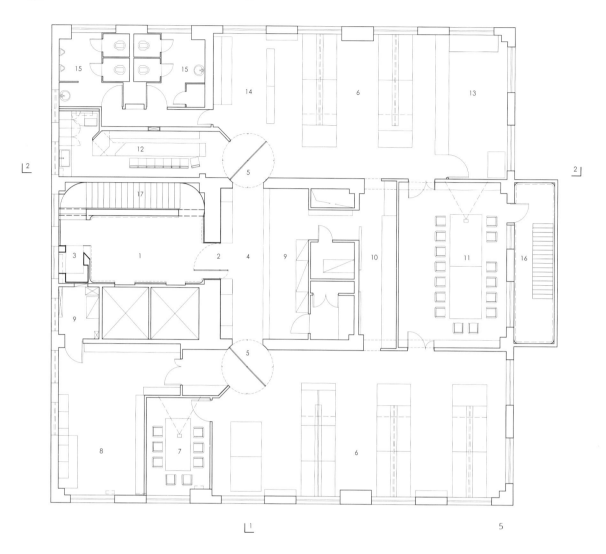

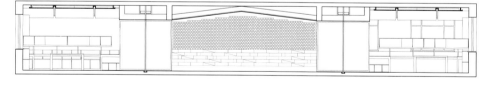

1 电梯厅 / Elevator bank
2 大门 / Gate
3 吸烟处 / Smoking pavilion
4 展廊 / Gallery
5 旋转镜门 / Mirror-door
6 设计工坊 / Design workshop
7 小会议室 / Small meeting room
8 办公室 / Office
9 储藏间 / Storage
10 打印走廊 / Print room
11 会议 / 图书室 / Meeting room
12 餐吧 / Bar
13 模型制作间 / Model making
14 材料样品角 / Material corner
15 卫生间 / Lavatory
16 露台 / Terrace
17 楼梯 / Staircase

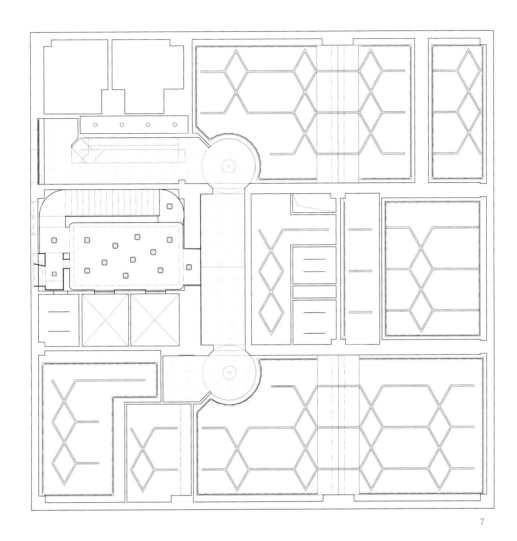

7

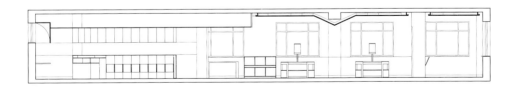

8

10

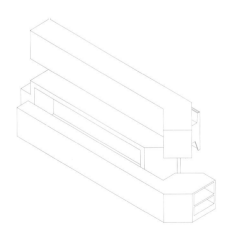
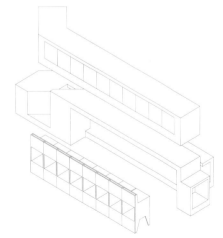

11

12

13

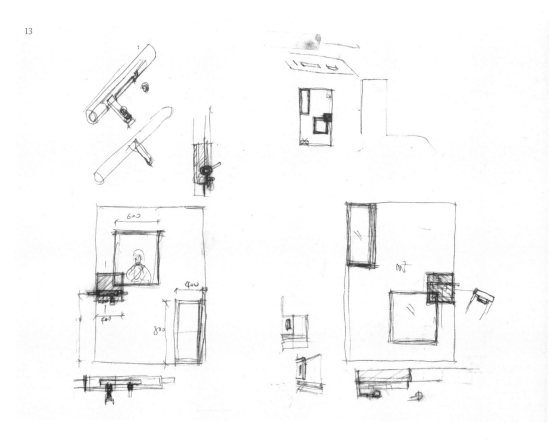

12 大门局部 / Detail of the gate
13 大门草图 / Sketch of the gate
14 电梯厅 / View of the elevator bank

14

2010

建川镜鉴博物馆暨汶川地震纪念馆
JIANCHUAN MIRROR MUSEUM & WENCHUAN EARTHQUAKE MEMORIAL

　　建川镜鉴博物馆暨汶川地震纪念馆设计始于 2004 年 2 月，2010 年 9 月建成投入使用，位于四川大邑安仁古镇，是民间投资建设的建川博物馆聚落中的单体建筑之一，由博物馆及其商业街坊混合而成。最初设计为镜鉴博物馆，收藏展示"文革"时期的各种镜面。汶川地震发生后，通过设计改造增加了地震纪念馆，以"震撼日记"的形式收藏和展示地震文物和相关艺术创作，成为一个两馆一体的复合型博物馆。建筑外部平和而宁静，却混合收藏着内部的虚像狂乱和现实震荡，给予当代人对过往历史和灾难的即时体验，加上其曲折、多变、富于现实感和戏剧性的设计建造历程，可称得上是这个时代的"镜鉴"之馆。

　　沿着周边街道优先安排的商业单元（底层商铺＋上层住宅）围合着内部的大小庭院和"复廊"式展览空间及其风车状线型延伸的博物馆建筑形态。原本设计建造并停工难产的镜鉴馆由于汶川地震的发生加入了地震纪念馆而再造重生，经设计改造后的两个馆在空间上相互叠加又各自独立并置，"虚像"和"现实"相互混合、对照。镜鉴馆内部空间通体采用白色喷漆花纹钢板，通过分布于"复廊"交会节点处的不同方式组合的可旋转"镜门"装置，造成纯净、抽象、变幻多端的虚像空间，而让参观者体验到那种极端的狂热和失序；地震馆则保留原始的混凝土墙浇注和砖墙砌筑痕迹，只作局部简单刷白，以临时、粗粝、具体、真实的空间和展品让参观者感受痛切而震撼的现实。两者以各自的方式纪念、展示并让参观者能体验那两段历史，给予后人以鉴戒、警示和启迪。

　　主要外墙材料中，清水混凝土用在沿街防火墙，红砖用在沿街的商铺外墙，暗示"红色年代"，而青砖则用于庭院的建筑外墙，对应内向空间的静谧深沉。当地丰富的砌砖传统在这里体现为在同一砖砌模数单元控制下不同通透程度的砖砌花墙，以满足采光、通风、景观、私密性和其他功能性要求。建筑师为此专门设计了透明"钢板玻璃砖"，造价低廉并易于加工，用于花墙上对应室内空间的砌空部分。

摄影 / Photographer:
张广源 / Zhang Guangyuan

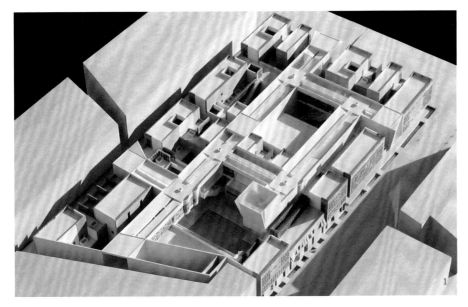

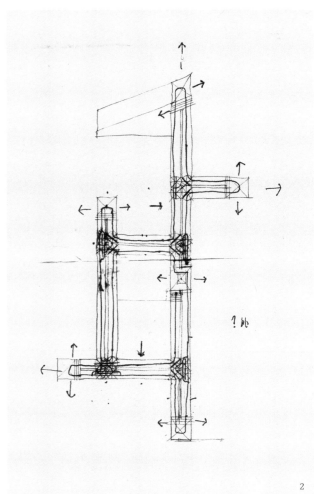

The Jianchuan Mirror Museum & Wenchuan Earthquake Memorial, located at the ancient town of Anren in Dayi County, Sichuan Province, was originally designed in February 2004 and put into operation in September 2010. It is one of the buildings among the non-governmentally-funded Jianchuan museum cluster, consisting of a mixture of museums and commercial streets. The museum was originally designed to collect and exhibit mirrors used during the Chinese Cultural Revolution. After the Wenchuan Earthquake, it was redesigned into a combined museum which added Wenchuan Earthquake relics and the relevant artworks memorializing the event in the form of a "Shock Diary". Unlike its peaceful and tranquil exterior, the interior displays disorienting images allowing people to experience and memorialize the disasters. With its dramatic design and construction, its interior twists and turns, and fluctuations between reality and illusion, the museum complex is a contemporary "Hall of Mirrors".

The commercial areas (ground-floor shopping area and upper-level residences) framed by the surrounding streets enclose the museum's large and small internal courtyards. With the courtyard as the "hub," the long, "double gallery" exhibition space extend out from the courtyard in a windmill shape. The museum complex, which was "reborn" after the earthquake, were renovated to juxtapose the "virtual" with the "real". The interior of the Mirror Museum is made of white painted checkered steel plates. We added rotating mirror doors at "double gallery" corner intersections to create an ever-changing, disorienting environment to stimulate the frenzy and chaos of the Cultural Revolution. In contrast, with partial, uncomplicated whitewashing, the Earthquake Memorial retains traces of its concrete wall casting and bricks. This makeshift, rough, cold concrete space immerses visitors in a feeling of pain and trauma. The museum and memorial enable visitors to experience two great human tragedies —the manmade disaster of the Cultural Revolution and the natural disaster of the Wenchuan Earthquake—of differing origins that teach later generations with warning and alert.

The exterior wall is formed with pale concrete, red and grey shale bricks. Concrete was used for a fire wall along the street. Red bricks were used for the exterior wall of the shops along the street, indicating the "Red Era". Grey bricks were used for the exterior wall of the courtyard, appropriate for a quiet and introspective space. A rich bricklaying tradition is on display on the brick walls, where the bricks and masonry can be arranged for transparency and modularity. The opacity of the brickwork varies depending on light, ventilation, ambience, and degree of privacy necessary for any given indoor use. For the museum and memorial, we used low cost, easy to process "glass bricks" to fill in voids in the brickwork to protect the interior.

4 平面草图 / Sketch of the plan
5 游赏路径草图 / Sketch of the exhibition route
6-7 构造草图 / Detail sketches

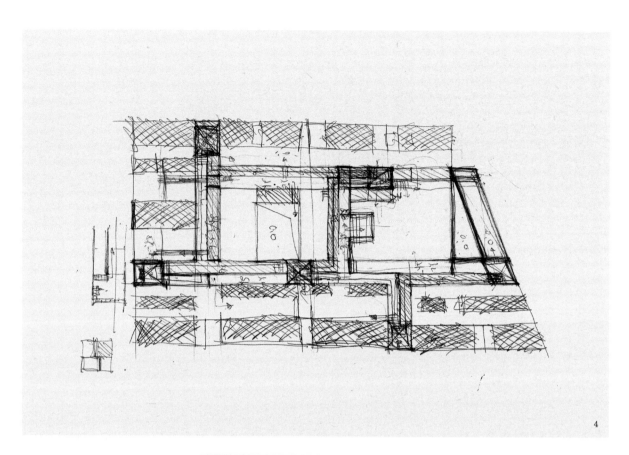

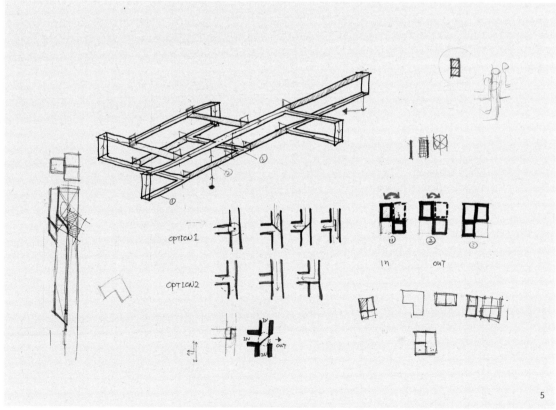

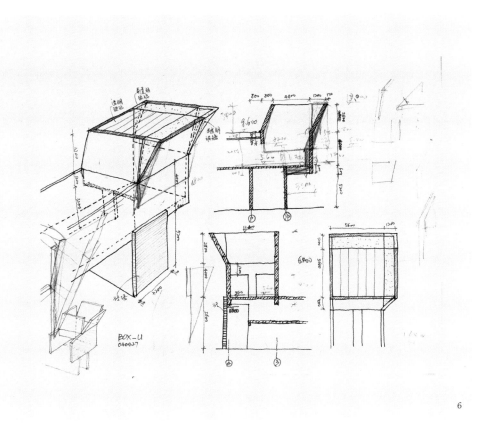

BOX-U
040427

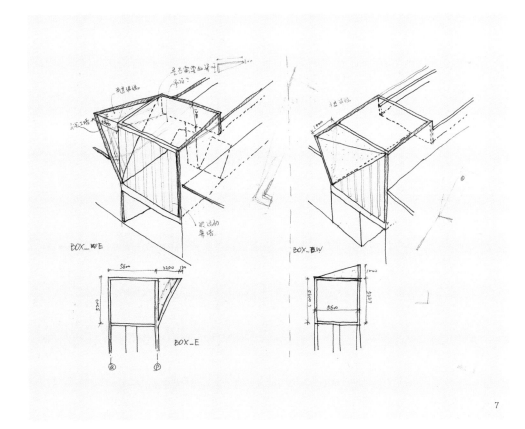

BOX_NVE

BOX_E

BOX_BW

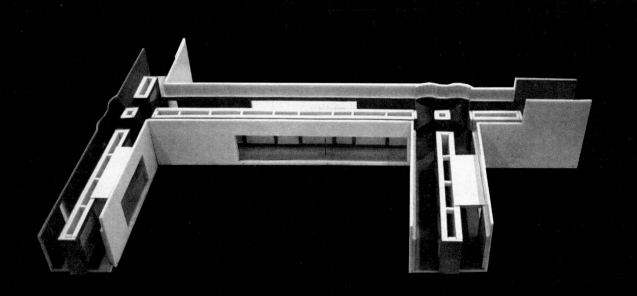

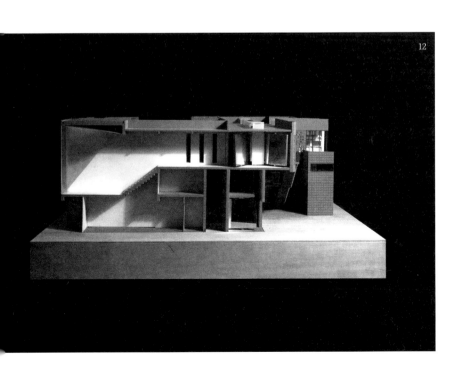

13

0 5 10   30m

14

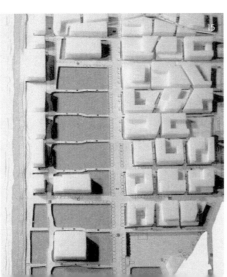

15

16

17

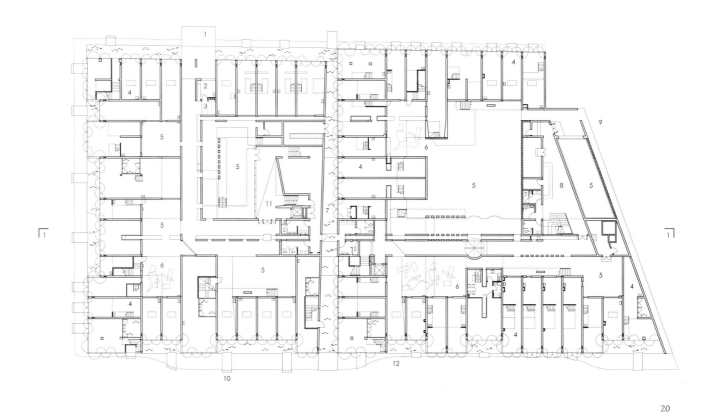

20

21

1 地震馆入口 / Earthquake Memorial entrance
2 纪念品商店 / Shop
3 售票处 / Ticket
4 艺术家工作室 / Workshop
5 庭院 / Courtyard
6 室外展场 / Outdoor exhibition
7 街巷 / Alley
8 结束厅 / Ending hall

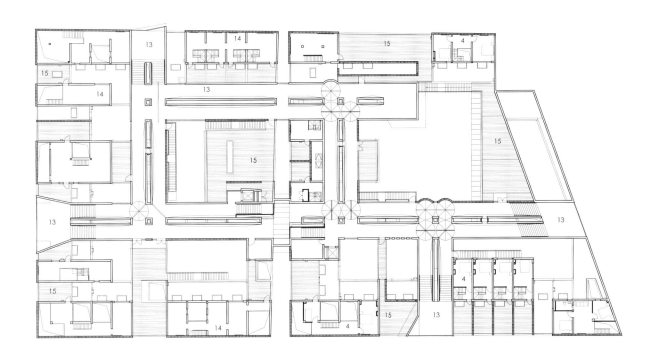

22

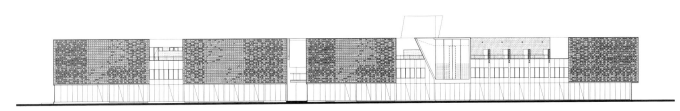

23

24 不同尺寸的钢板玻璃砖 / Bricks made of steel plate and glass in different sizes
25 施工中的砖墙 / The brick wall under construction
26 花墙详图 / Details of the wall
27 主庭院人视 / View of the main courtyard

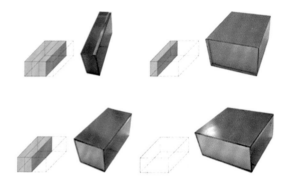

24

25

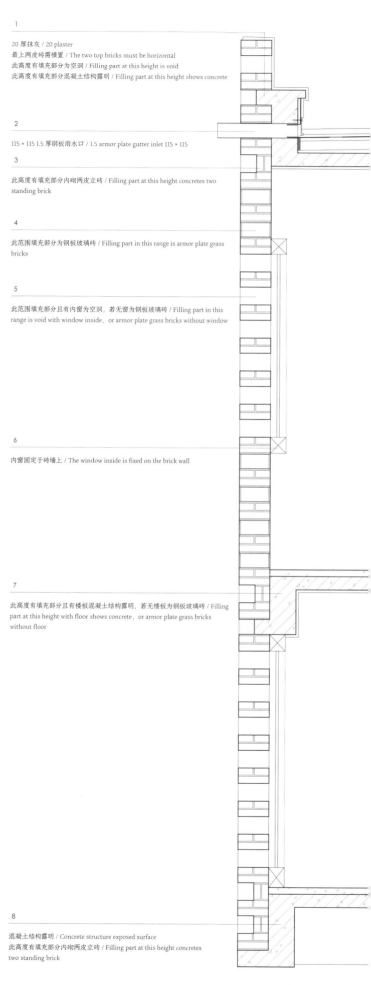

1

20 厚抹灰 / 20 plaster
最上两皮砖需横置 / The two top bricks must be horizontal
此高度有填充部分为空洞 / Filling part at this height is void
此高度有填充部分混凝土结构露明 / Filling part at this height shows concrete

2

115 × 115 1.5 厚钢板雨水口 / 1.5 armor plate gutter inlet 115 × 115

3

此高度有填充部分内砌两皮立砖 / Filling part at this height concretes two standing brick

4

此范围填充部分为钢板玻璃砖 / Filling part in this range is armor plate grass bricks

5

此范围填充部分且有内窗为空洞，若无窗为钢板玻璃砖 / Filling part in this range is void with window inside，or armor plate grass bricks without window

6

内窗固定于砖墙上 / The window inside is fixed on the brick wall

7

此高度有填充部分且有楼板混凝土结构露明，若无楼板为钢板玻璃砖 / Filling part at this height with floor shows concrete，or armor plate grass bricks without floor

8

混凝土结构露明 / Concrete structure exposed surface
此高度有填充部分内砌两皮立砖 / Filling part at this height concretes two standing brick

26

27

2010

北京地铁 4 号线及大兴线地面
出入口及附属设施
ENTRANCES FOR
LINE 4 & DAXING LINE OF
THE BEIJING SUBWAY

北京地铁 4 号线及大兴线地面出入口及附属设施项目设计分别始于 2008 年 8 月（4 号线）和 2010 年 2 月（大兴线），分别于 2009 年 10 月（4 号线）和 2010 年 12 月（大兴线）建成投入使用，作为北京西部城区贯穿南北的地铁大动脉的地面出入口设施，使用和运营状况良好，交通识别性强，并已成为地铁沿线的特色城市景观。

地铁 4 号线是北京市轨道交通网中由南至北穿越了新旧北京城区的轨道交通线，线路全长 28.177 公里，全程设 24 座车站。其地面出入口及附属设施的设计需要面对城市空间的特殊性、地面设施类型的复杂性、地下预留站体的结构多变性以及紧张的工期等诸多前提条件。利用钢结构便于标准化、模数化、预制化的特点，将全线出入口站亭设计为在一定模数控制下 4 种不同规格的系列化网格状钢结构，以适应复杂的地下站体结构尺寸，并在工厂预制标准化的钢结构网格以及与之对应的外墙板块单元，进行现场拼装。在外墙预制板块设计中引入"城市画框"的概念，运用金属板、彩釉印刷玻璃和透明玻璃的不同组合，对变化的城市环境进行多样性摄取，使用者可以透过取景窗辨识出入口所处城市空间的典型特征。建筑形体采用了坡形山墙断面和矩形断面过渡的基本形式，凸显地铁出入口的功能性与标识性，同时呼应了 4 号线串联旧城和新城的文脉关系，使地铁出入口站亭成为市民了解城市文化，体验城市空间，从小建筑中见大意境的公共交通建筑。

地铁大兴线作为 4 号线的南延线，处于由现代城市景观向郊区自然景观过渡的特定城市空间之中，其地面出入口及附属设施依然延续了 4 号线坡形断面与矩形断面过渡的基本形式。为了突出本线的地理文化特征，引入了"城市画卷"的设计概念，运用现代彩釉玻璃丝网印刷技术，将古代山水画作经过抽象后拓印于玻璃幕墙之上，唤起出入站亭的乘客对大兴地区自然景观的历史记忆和与现代时空交织的独特体验。整幅画面采用 5 种不同透明度的白色釉块组合抽象表现，虚实相叠，远观其势，近观其质，并利用白天自然光线和夜晚室内光线的变化，形成对幕墙画面及城市景观多角度、多时段的不同解读。

摄影 / Photographer:
张广源 / Zhang Guangyuan

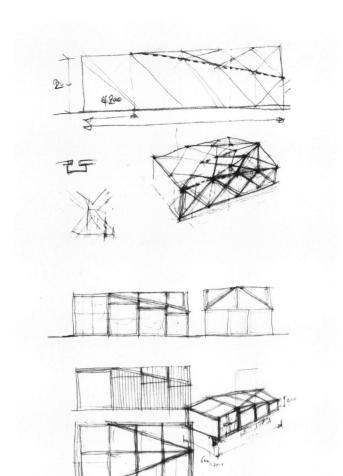

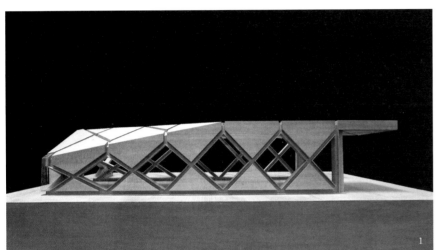

The Entrances for Line 4 & Daxing Line of the Beijing Subway were initially designed in August 2008 (Line 4) and October 2010 (Daxing Line), and completed and put into use in October 2009 (Line 4) and December 2010 (Daxing Line). These subway stations serve the the north-south artery along Beijing's west side. The facilities serve their utilitarian and operation functions and are easily identifiable at street level. These stations' entrance canopies have become familiar urban landmarks.

Subway Line 4 runs through a 28.177 km-long route connecting 24 stations from the south to the north along the new and old urban areas of the city. The architecture of its station entrances and facilities must meet many design specifications and construction considerations, such as the particularities of urban space near each station, the complex infrastructure aboveground, the varying structural requirements underground, and the tight construction schedule. Thus, all the entrances must adopt a steel lattice construction that can be easily standardized, modularized, and prefabricated. Four kinds of steel and glass template were designed that could be adapted depending on station's surrounding environment with the prefabricated steel frame and glass plates assembled on site. The design of the station entrances introduced the concept of the "urban picture frame". Different combinations of metal plates, glazed glass, and clear glass captured the different urban surroundings of each station. The entrance adopts the form of a sloped gable that transitions to a rectangular cross section. These structural elements identify their function as well as highlight Line 4's role as gateways connecting the old to the new areas of the city. The entrances guide citizens into and out of the stations through the history and culture of the city.

At the southern extension of Line 4, the Subway Daxing Line is located in a special space where the modern city transitions to a more natural suburban landscape. The entrance designs adopt a formal language similar to that of Line 4 station entrances. We introduced the "urban scroll painting" concept to highlight this route segment's geographical and cultural features. Using modern glazed glass screen printing technology, abstracted ancient landscape paintings were printed on the glass canopies to evoke the passengers' unique experience of the traveling though modern time and space while remembering the natural landscapes that had developed since antiquity in the Daxing area. By glazing the glass with different levels of opacity, people are presented with a combination of abstract expressions of the environment surrounding the station. Depending on the viewing perspective, the movement of the sun during the day, and the projection of interior artificial lights at night, the setting varies through the glass curtain and combines real and virtual elements. Passengers will be attracted by the changing light and shadow from the glass curtain and will experience the interaction between the city's past and present.

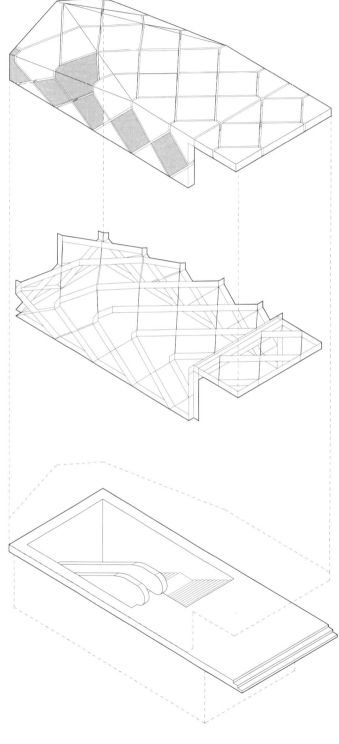

13

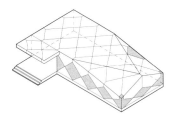
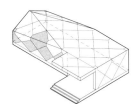
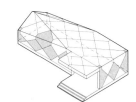
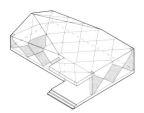

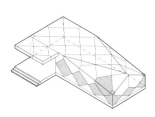
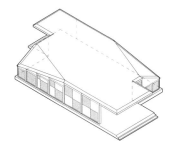
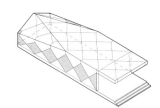
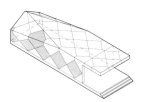

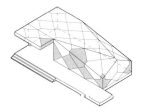
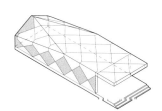
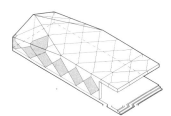
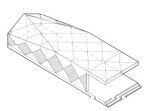

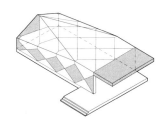
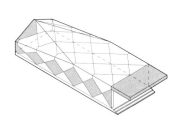
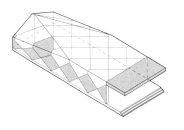

14

15

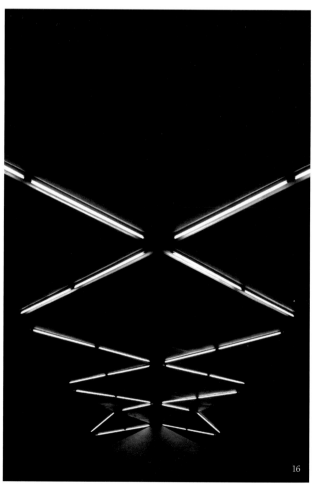

16

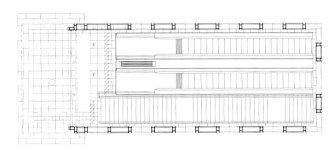

17

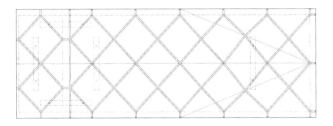

18

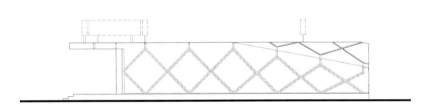

19

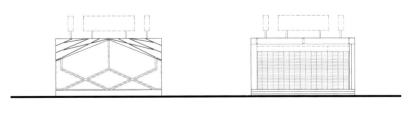

20 21

24

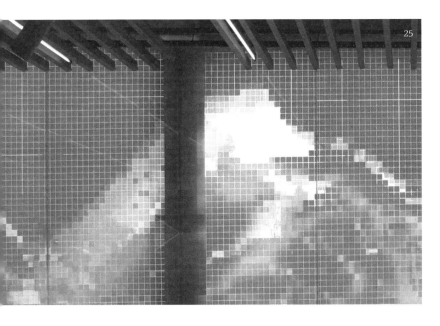

25

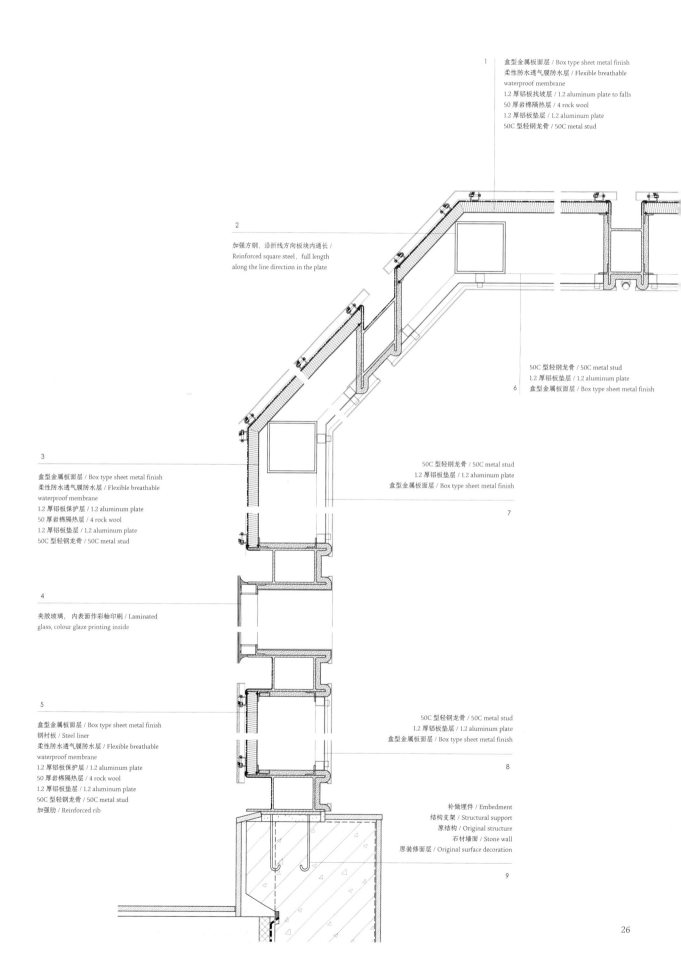

1

盒型金属板面层 / Box type sheet metal finish
柔性防水透气膜防水层 / Flexible breathable
waterproof membrane
1.2 厚铝板找坡层 / 1.2 aluminum plate to falls
50 厚岩棉隔热层 / 4 rock wool
1.2 厚铝板垫层 / 1.2 aluminum plate
50C 型轻钢龙骨 / 50C metal stud

2

加强方钢，沿折线方向板块内通长 /
Reinforced square steel，full length
along the line direction in the plate

50C 型轻钢龙骨 / 50C metal stud
1.2 厚铝板垫层 / 1.2 aluminum plate
6 盒型金属板面层 / Box type sheet metal finish

3

盒型金属板面层 / Box type sheet metal finish
柔性防水透气膜防水层 / Flexible breathable
waterproof membrane
1.2 厚铝板保护层 / 1.2 aluminum plate
50 厚岩棉隔热层 / 4 rock wool
1.2 厚铝板垫层 / 1.2 aluminum plate
50C 型轻钢龙骨 / 50C metal stud

50C 型轻钢龙骨 / 50C metal stud
1.2 厚铝板垫层 / 1.2 aluminum plate
盒型金属板面层 / Box type sheet metal finish

7

4

夹胶玻璃，内表面作彩釉印刷 / Laminated
glass, colour glaze printing inside

5

盒型金属板面层 / Box type sheet metal finish
钢衬板 / Steel liner
柔性防水透气膜防水层 / Flexible breathable
waterproof membrane
1.2 厚铝板保护层 / 1.2 aluminum plate
50 厚岩棉隔热层 / 4 rock wool
1.2 厚铝板垫层 / 1.2 aluminum plate
50C 型轻钢龙骨 / 50C metal stud
加强肋 / Reinforced rib

50C 型轻钢龙骨 / 50C metal stud
1.2 厚铝板垫层 / 1.2 aluminum plate
盒型金属板面层 / Box type sheet metal finish

8

补做埋件 / Embedment
结构支架 / Structural support
原结构 / Original structure
石材墙面 / Stone wall
原装修面层 / Original surface decoration

9

26

2010

北京地铁昌平线西二旗站
XI'ERQI STATION
OF CHANGPING LINE
OF THE BEIJING SUBWAY

　　北京地铁昌平线西二旗站设计始于 2008 年 11 月，2010 年 12 月建成投入使用，是北京地铁昌平线的南起 / 终点站，也是与已建成运行的城铁 13 号线的换乘站，起 / 终点站 + 换乘车站复杂程度高，原有的 13 号线客流量极大，并且在新的换乘车站建设期间，原有的 13 号线西二旗车站要同时正常运行，种种因素使得这个车站的设计具有前所未有的挑战性。

　　确保人们使用方便、顺畅、安全，是地铁车站设计的基本原则。经反复研究最终确定的车站形式为半地下一层、地上二层，四柱三跨框架式高架车站，其中的昌平线部分为高架侧式站台（二层），城铁 13 号线部分为地面侧式站台（一层），站厅位于地面一层，根据侧式站台的特征，车站采用双四边形组合建筑断面，并以 PTFE 膜结构作为屋面和立面维护材料和结构，实现了内部基本无柱的长向大空间。使用电脑模拟的手段来检验和修正车站运行及换乘系统，以实现极端条件下地铁客流交通的顺畅性和安全性。

　　适应地铁交通需求的双四边形组合筒状空间被直接外现为简洁、清晰、可读的建筑造型，并采用了折纸状的结构和形式，实现了模数化、标准化、预制化的设计与建造。折纸暗示了一种很轻的、空间上有韵律感的结构和材料，半透明的膜材跟纸有一种对应性，膜材本身也有一定结构作用，其自身张力和钢结构之间形成一种结构的复合作用，膜材像折纸一样的转折在光线作用下让人感受到奇妙的雕塑感，简洁、流畅而富于韵律，真实而充分地体现了张拉膜材料和结构的特征，建筑形式也与结构、排水等要求相适应。车站的出入口大厅、雨篷等元素采用了一体化的建筑语汇。

　　由于 PTFE 膜材料的半透光特性，白天的车站内可以透射进柔和而充分的自然光线，无须人工照明；到了夜晚，车站内的灯光透射出去，使得车站就像两个发光的纸灯笼，前来乘坐地铁的人远远就可以看到，城市中这个不大的建筑带给人回家的温暖感。地铁西二旗站以其"弱"和"轻"的建造方式和不动声色的建筑姿态与人们的日常生活密切关联，并成为重要的城市标志物。

摄影 / Photographer:
张广源、黄达达 / Zhang Guangyuan, Huang Dada

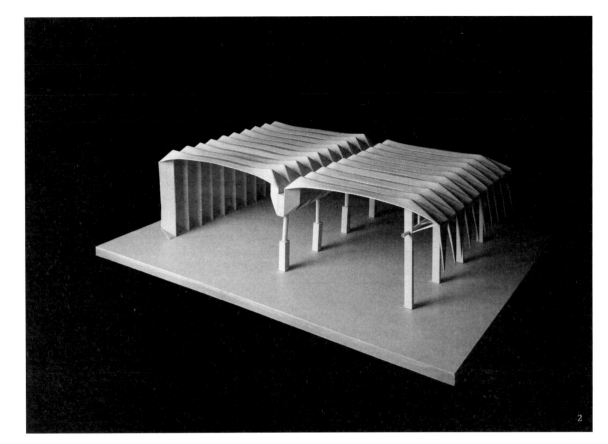

1

2

Xi'erqi Station, the southern terminus of the Changping Line of the Beijing Subway was initially designed in November 2008 and put into operation in December 2010 as the transfer station between Beijing Metro Line 13 and the Changping Line. As a terminal and transfer point, the Xi'erqi Station was already a complex design, but we also needed to consider the high passenger flow on the existing Line 13. Because Line 13 continued its normal operation during construction of the new transfer station, this project faced additional, unique challenges.

The basic principle of subway station design is to ensure that passengers experience convenient, efficient, and safe movement. After extensive research and development, we designed the station as a structure with a concrete, elevated frame. The station consisted of a partially subterranean first floor with two stories above-ground supported by four colonnades creating three spans. The Changping Line is located on an elevated platform on the second story. Line 13 is part of a platform on the ground story. Due to the characteristics of side-platforms, the cross-section of the buildings consists of two quadrangles combined with each other. The PTFE membrane serves aesthetic and structural purposes as it encloses the roof and facade to provide an open space without structural columns blocking sightlines or flow. Computer simulations allowed planners to modify the existing operations and transfer system to ensure continued efficiency and safety of movement under extremely busy conditions.

The dual rectangular tunnels adapted to facilitate the subway flow are directly presented as a simple, clear and readable architectural form. This "origami" structure are adopted to be prefabricated, standardized, and modular in design and construction. The structure's resemblance to origami bestows the structure with a sense of lightness and rhythm. The translucent membrane appears as light as paper, but still gives the structure strength. The PTFE membrane creates a tension with the steel structure. Under the origami folds of the membrane, passengers feel as if they are in the midst of a light sculpture. The inherent characteristics of the membrane can be taut and fluid, creating a rhythmic architecture. Additionally, this design is compatible with the drainage needs of the station. The entrance hall, awning and other elements of the station adopt a range of architectural vocabulary.

Because of the translucence of the PTFE membrane, soft and sufficient natural light can be transmitted into the station without artificial lighting during the day. At night, the lamps transmit light from the building allowing the station glow like two paper lanterns that can be seen from far away by subway commuters. This small building in the city gives people going home every day a sense of warmth. Because the station is constructed using "delicate" and "light" materials, it strikes a calming architectural relationship with the daily lives of commuters and has become an important city landmark.

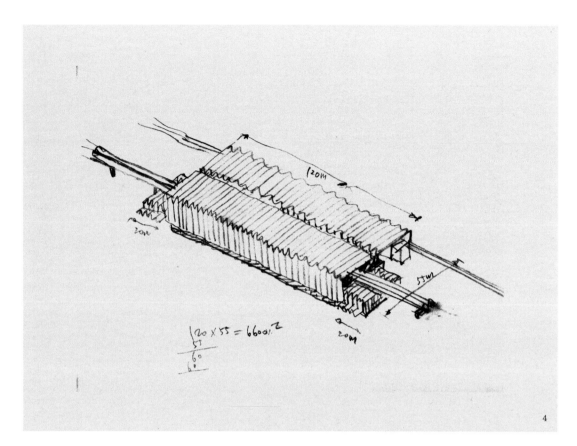

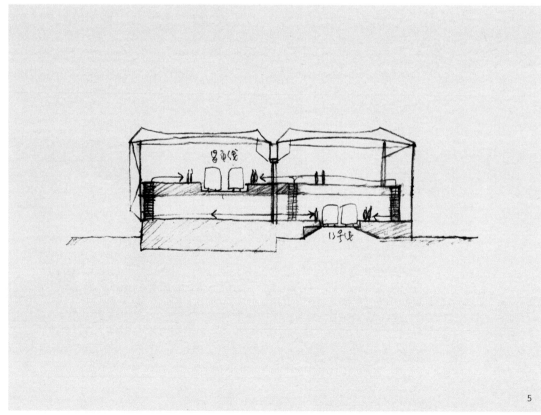

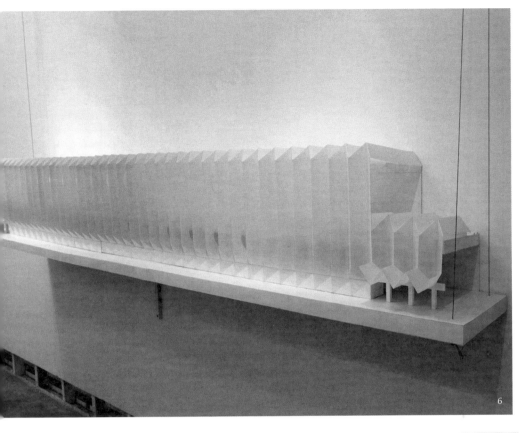

8 西南侧鸟瞰 / Aerial view from the southwest
9 总平面图 / Site plan
10 西北侧人视 / View from the northwest

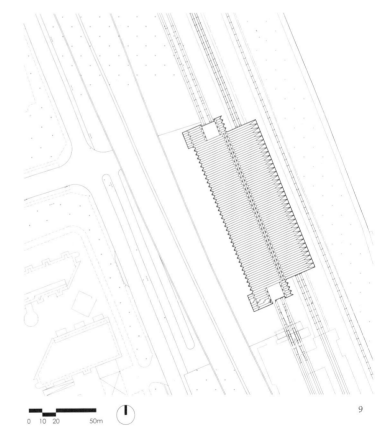

0 10 20   50m

9

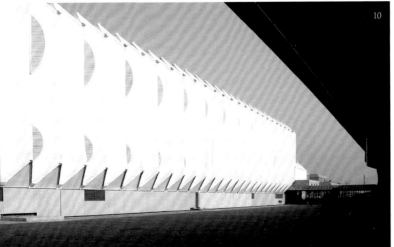

10

8

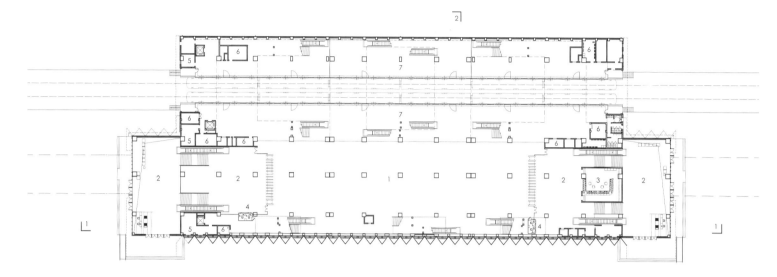

11

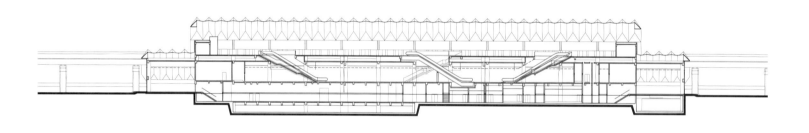

12

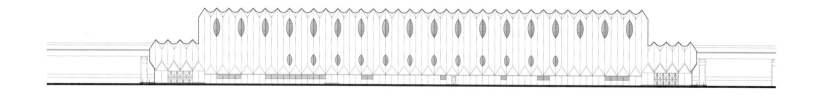

13

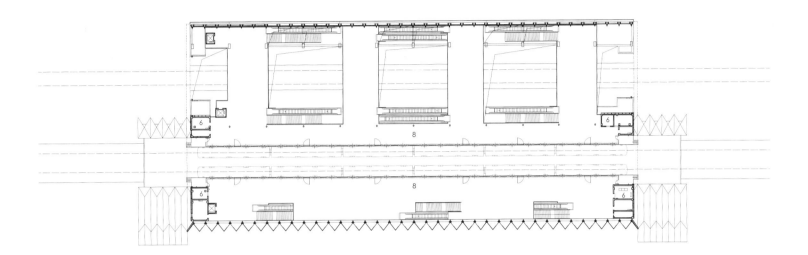

14

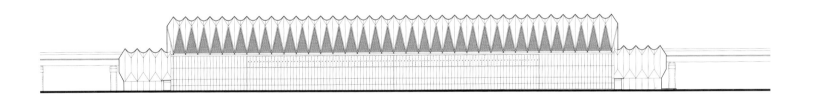

15

0 2   5   10m

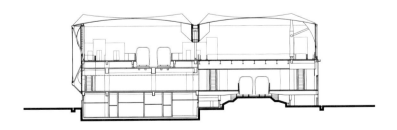

16

1 付费区 / Paid area

2 非付费区 / Unpaid area

3 合用车站控制室 / Shared station control room

4 客服中心 / Passenger centre

5 办公 / Office

6 设备用房 / Equipments

7 站台（13 号线）/ Platform of Line 13

8 站台（昌平线）/ Platform of Changping Line

17

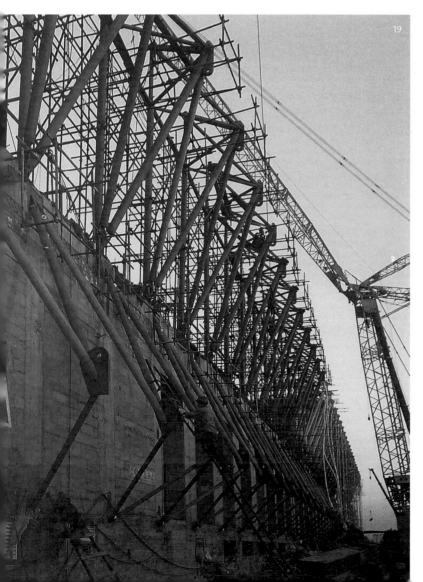

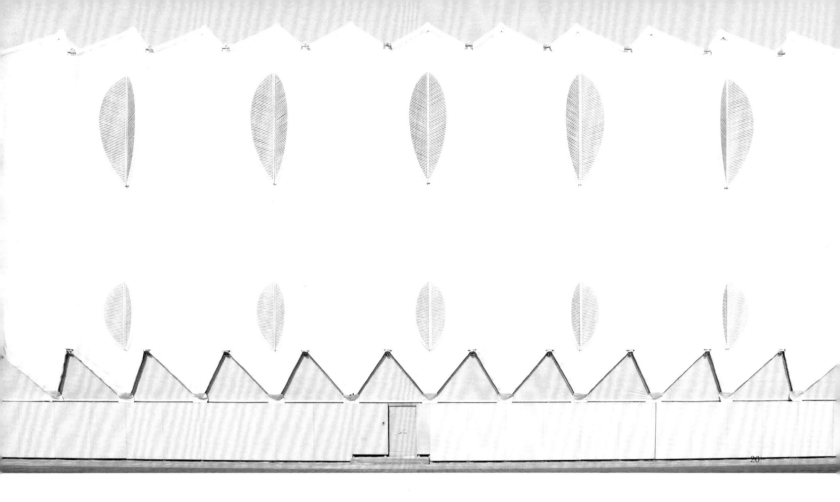

20

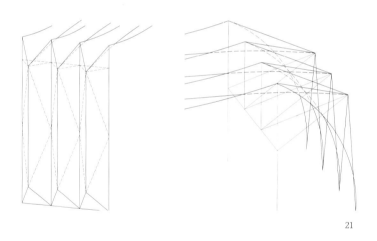

21

2011

海南国际会展中心
HAINAN INTERNATIONAL CONVENTION & EXHIBITION CENTER

海南国际会展中心设计始于 2009 年 7 月，2011 年 7 月建成投入使用，位于海口市西部，新的城市组团北部，城市南北向景观轴线的尽端；用地北侧直面琼州海峡，周围与酒店、公寓、城市公园等相邻，包括展览中心和会议中心两大功能区及其相关附属设施。

不同于把超大体量建筑形体分解、分散布置的常规做法，建筑主体在整体布局和空间构成上采用一体化的设计手法，将展览中心和会议中心"聚零为整"，整合处理为一个巨大的完型体量，匍匐在南中国海滨，成为城市与大海之间的联结物和南北向景观轴线的结束与高潮。连续的外轮廓将各部分建筑功能区统合在一起，彼此联结，并呈现出扩张与收缩、凸显与内敛的平衡。作为新城轴线的底景，这一庞大的建筑有着明确的、与新城中心合一的中轴，同时又因地形、海岸和内部功能的不同而产生了丰富的非对称建筑形体，成为在规则韵律中富含变化、整体造型协调统一的一个巨型建筑体。建筑造型是处于"像与不像"之间的抽象人工物，"海水""云团""沙滩""海洋生物""海上景象"……似是而非，但总与大自然之海洋气质相契合。

屋顶中部隆起，坡向边缘，整个屋面分为正弦曲面的中央区域与直纹曲面的边缘区域，根据建筑内部功能的不同需要，屋面的隆起高度和波状起伏的剧烈程度有所不同，展览中心部分较平缓，会议中心部分较陡急。中央区域屋面以双向正弦曲线起伏波动，形成薄拱壳体，上凸正壳和下凹反壳的平面单元投影呈方形，交替相连为 22m × 22m 的基本展览单元，由管径 325mm 等截面的圆形钢管梁密格式布置而成，以工厂预制、现场组装、吊装就位的方式建造，大部分室内钢结构完全露明，呈现出结构自身的美感和韵律，壳顶单元的天窗和高侧窗为室内空间提供了充足的、不断变幻角度的自然光线，建筑外观与室内空间高度统一，充分体现出建筑形式、结构、空间的同一性特征。

环绕整个建筑外侧的半拱形檐廊、雨棚及其下结合窗台设计的"石凳"，是应对当地强烈的日晒条件，为室外的使用者提供的遮阳和休息设施。采用的建筑材料有石材、曲面铝板、现制 GRC 及屋面涂料等。

摄影 / Photographer:
张广源、李兴钢、张玉婷 / Zhang Guangyuan, Li Xinggang, Zhang Yuting

1 基本结构空间单元模型（1/50） / Model of the basic structure-space unit (1/50)
2 前期草图 / Concept sketch
3 总体模型（1/1000） / Model (1/1000)

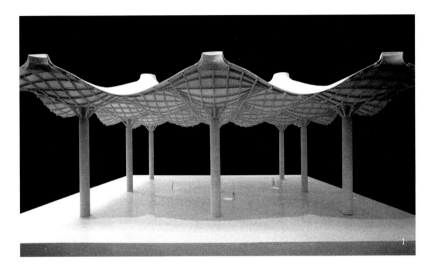

The Hainan International Convention & Exhibition Center was initially designed in July 2009 and completed and put into use in July 2011. It is located in western Haikou City, Hainan Province, in the north of the new city cluster, and at the end of the north-south axis of the city. Adjoining the Qiongzhou Strait on the north, its site is surrounded by hotels, apartments, and a city park. The center provides exhibition and conference space and related support facilities.

Unlike the conventional method of breaking down and disaggregating large buildings, the central body of the building adopts the integrated design method in its overall layout and spatial composition. The exhibition and conference centers are combined into a large, unified mass that connects the city to the sea. It is the terminal point and climax of the north-south axis of the city. The continuous, undulating roof integrates the building's various functional areas together, connecting, and showcasing the balance between expansion and contraction, exuberance and restraint. As the foreground of city axis, this huge building has a distinct axis that coincides with the center of the new urban area. To accommodate the location's terrain, the center's relationship to the coastline, and its diverse internal usage requirements, we needed to design multiple, asymmetrical, yet still rhythmical architectural shapes for the structure. It is a massive building with a rich architectural vocabulary, but the center retains a harmonious overall shape. The center is our attempt at a manmade abstraction that includes the sea, clouds, the beach, and marine creatures that mimics the diverse coastal ecosystem where it resides.

The roof gently slopes up from its edge into a large, flat plateau with an undulating surface of sinusoidal waves. The roof height and the steepness of its undulations change in accordance with the spatial functions of the corresponding interior beneath it. The exhibition center roof is relatively flat, while the conference center roof is relatively steep. The roof in the central area fluctuates in a two-way sine curve, forming a thin arch shell. The area for each exhibition unit measures 22 meters × 22 meters and above each unit is a concave and convex ceiling. The circular steel tube beams with the diameter of 325 millimeters are arranged in a dense grid. They are prefabricated at the factory, assembled on-site, and hoisted into place. Most of the indoor steel structures are completely exposed to feature its elegance and rhythm. The skylight and portholes placed high along the sides of the upper shell provide sufficient and constantly changing natural light for the interior space. The building's appearance is highly unified with the interior space, fully reflecting the identity of building's form, structure and space.

All around the exterior of the building are semi-arch eaves and canopies, as well as stone benches designed in line with windowsills. The architecture has been built with stone, cambered aluminum plates, cast-in-situ GRC, and roof coating.

5 体量模型（1/2000）/ Volume models (1/2000)
6 城市设计草图 / Sketch of the urban design
7 平面草图 / Plan sketch
8 立面草图 / Elevation sketch

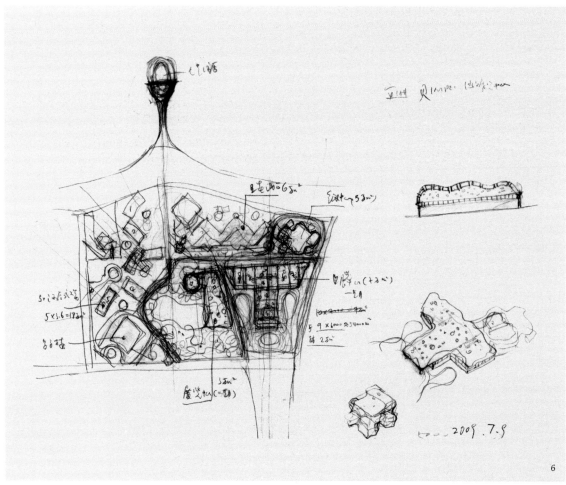

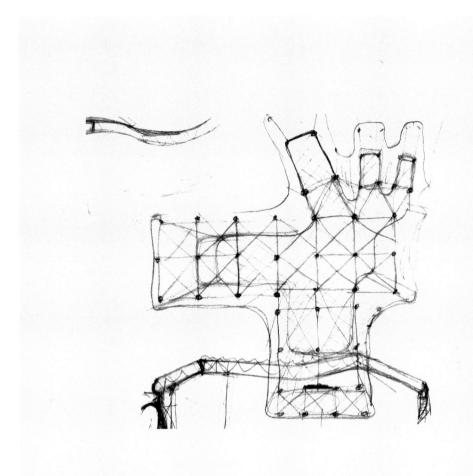

7

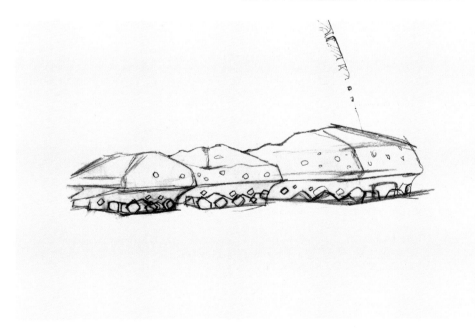

8

9

10

0 50 100   200m

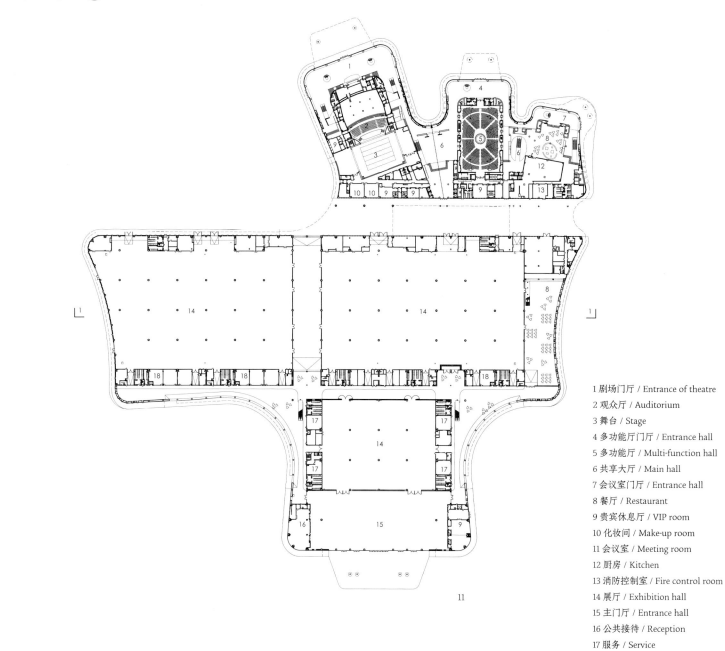

1 剧场门厅 / Entrance of theatre
2 观众厅 / Auditorium
3 舞台 / Stage
4 多功能厅门厅 / Entrance hall
5 多功能厅 / Multi-function hall
6 共享大厅 / Main hall
7 会议室门厅 / Entrance hall
8 餐厅 / Restaurant
9 贵宾休息厅 / VIP room
10 化妆间 / Make-up room
11 会议室 / Meeting room
12 厨房 / Kitchen
13 消防控制室 / Fire control room
14 展厅 / Exhibition hall
15 主门厅 / Entrance hall
16 公共接待 / Reception
17 服务 / Service
18 商店 / Shop

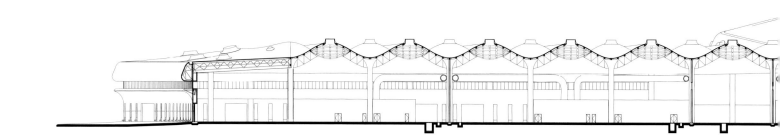

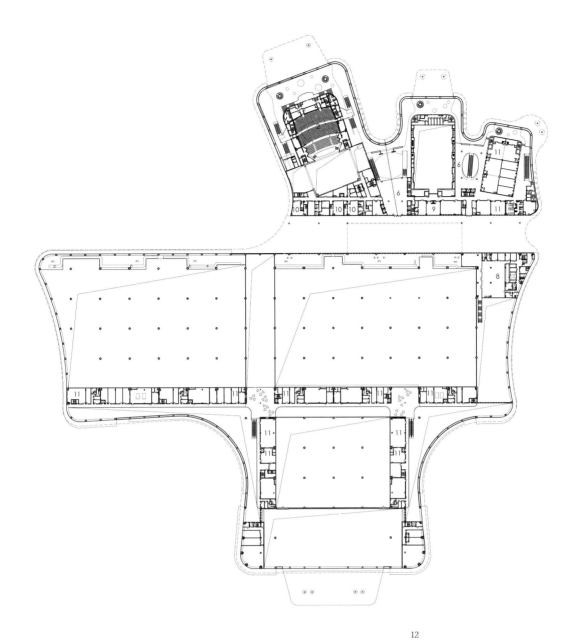

12

13

15

14

16 展厅之间的廊道空间 / View of the gallery space between exhibition halls
17 展厅内景 / Interior view of the exhibition hall

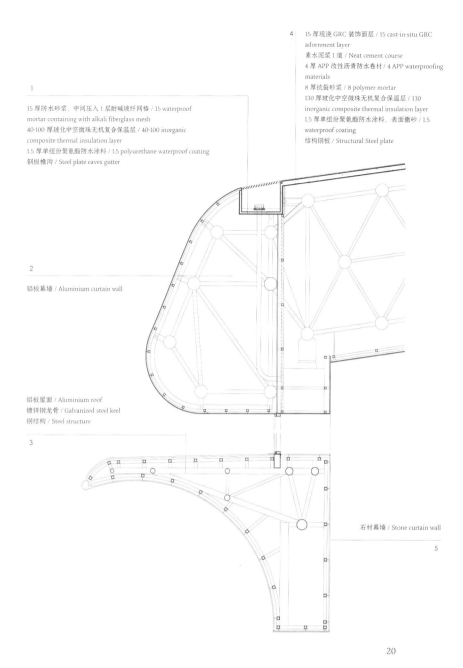

4 15 厚现浇 GRC 装饰面层 / 15 cast-in-situ GRC
   adornment layer
   素水泥浆 1 道 / Neat cement course
   4 厚 APP 改性沥青防水卷材 / 4 APP waterproofing
   materials
   8 厚抗裂砂浆 / 8 polymer mortar
   130 厚玻化中空微珠无机复合保温层 / 130
   inorganic composite thermal insulation layer
   1.5 厚单组份聚氨酯防水涂料,表面撒砂 / 1.5
   waterproof coating
   结构钢板 / Structural Steel plate

1 15 厚防水砂浆,中间压入 1 层耐碱玻纤网格 / 15 waterproof
   mortar containing with alkali fiberglass mesh
   40-100 厚玻化中空微珠无机复合保温层 / 40-100 inorganic
   composite thermal insulation layer
   1.5 厚单组份聚氨酯防水涂料 / 1.5 polyurethane waterproof coating
   钢板檐沟 / Steel plate eaves gutter

2 铝板幕墙 / Aluminium curtain wall

   铝板屋面 / Aluminium roof
   镀锌钢龙骨 / Galvanized steel keel
   钢结构 / Steel structure

3

   石材幕墙 / Stone curtain wall

5

20

4 15 厚现浇 GRC 装饰面层,表面做无色保护剂 / 15 cast-in-situ GRC adornment
   layer, brush colorless protectants
   素水泥浆 1 道 / Neat cement course
   4 厚 APP 改性沥青防水卷材 / 4 APP waterproofing materials
   8 厚抗裂砂浆,中间压入 1 层耐碱玻纤网格 / 8 polymer mortar containing with
   alkali fiberglass mesh
   130 厚玻化中空微珠无机复合保温层 / 130 inorganic composite thermal
   insulation layer
   1.5 厚单组份聚氨酯防水涂料,表面撒砂 / 1.5 polyurethane waterproof coating
   结构钢板 / Structural steel plate

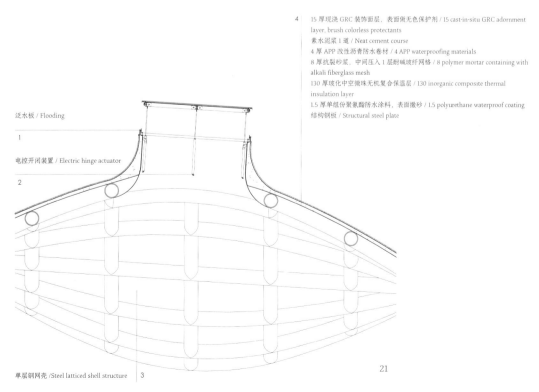

泛水板 / Flooding

1

电控开闭装置 / Electric hinge actuator

2

单层钢网壳 /Steel latticed shell structure   3

21

2011

元上都遗址工作站
VISITOR'S CENTER FOR SITE OF XANADU

元上都遗址工作站设计始于 2010 年 8 月，2011 年 8 月建成投入使用。元上都遗址是中国元代草原都城的遗迹，位于内蒙古自治区锡林郭勒盟正蓝旗五一牧场境内，山川雄固，草原漫漫。在遗址明德门之南约 1.5 公里的景区入口处设立工作站，解决遗址景区售票、警卫监控、管理办公、游客休息及公共卫生间等功能需求，并配合元上都遗址申报世界文化遗产。这个建筑以微小、轻盈、临时的自身存在感，表达对宏大、厚重、永固的草原和遗址环境的尊重。

建筑以化整为零的分散布局形成一个小小的草原聚落，偏于遗址轴线的一侧，既减小了完整大体量对环境的压迫，也留出了面向遗址的景观视线通廊。远远望去，一组白色坡顶的圆形和椭圆形小建筑，围合成对内和对外的两个敞开式庭院，分别供工作人员和游客使用。根据功能需求，这些小建筑大小不一、高低错落，相互之间的群体关系形成了有趣的对话。

由远及近，这组貌似却不同于通常蒙古包的草原建筑给造访的人们带来小小的戏剧性体验：圆形和椭圆形的建筑形体朝向庭院的部分，在几何体上被连续地切削，形成像建筑被剖开后连续展开的折线形内界面，原设计采用粗犷的清水混凝土做法，施工中被覆上了一层薄薄的白色涂料，仍然保留了施工木模板的痕迹；建筑形体朝向外侧的连续弧形界面，则罩以白色半透明的 PTFE 膜材，为建筑提供了一定的保温性能，又带来草原上"临时建筑"的轻盈感，膜与外墙之间的空隙里隐藏有结合膜结构撑杆固定的灯管，可在夜晚发出白色的微光，更显轻盈，似乎随时可以迁走一样，暗合草原的游牧特质，又最大限度地降低了对遗址环境的干扰；膜材和混凝土两种材料，交接于内庭院中由于弧形剖切形成的呈曲线形起伏波动的檐口轮廓线，构成了轻重、软硬的材质对比，同时，由转折而连续的墙面和屋檐所构成的具有强烈动感的人工界面，与苍茫的自然草原和静谧的遗址景观对话。混凝土外墙和天窗凸出于膜材表面。进入室内，结合弧墙和凸窗设计了固定式家具。透过外窗和天窗，光线进入房间，勾勒出方形的草原和天空的画面。

摄影 / Photographer:
张广源、李哲 / Zhang Guangyuan, Li Zhe

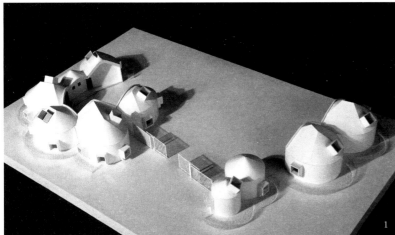

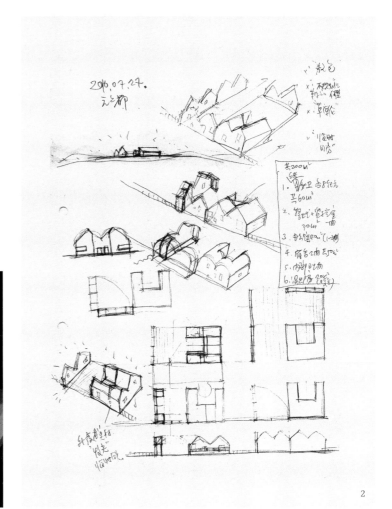

The Visitor's Center for Site of Xanadu was initially designed in August 2010 and put into operation in August 2011. The site of Xanadu—a relic of the grassland capital of the Yuan Dynasty—is located in the territory of Zhenglanqi, Inner Mongolia Autonomous Region, which is characterized by mountains, rivers and green grasslands. A tourism and information center is set up at the entrance of the scenic spot, approximately 1.5km south of Mingde Gate. The building serves the functional needs of site of Xanadu, including ticket offices, security, administrative offices, public restrooms for tourists and cooperate with the site to apply for World Cultural Heritage status. With a tiny, light and temporary sense of presence, this small settlement expresses its respect for the grand, deep and enduring grassland and site environment.

The building has a decentralized layout on one side of the site's axis, which not only reduces its volumetric tension with the environment, but also creates an uninterrupted sight-line to the site of Xanadu. Seeing from afar, two open courtyards for staff and visitors are surrounded by a group of small circular and oval buildings with sloped white facade. In accordance with functional requirements, these small buildings vary in size, height, and complexity. The relationships between the structures form an interesting dialogue.

As visitors approach the complex, some of the circular and oval buildings that they expect to be traditional, individual yurts are actually linked structures that unfold around the courtyard. The complex is made of rough concrete, which retains its wood grain image, is painted in a thin layer of white paint. The exterior's continuous curved surface is covered by white semi-translucent PTFE membrane. It provides certain thermal insulation properties and brings a sense of lightness like yurts erected on the grassland. Hidden in between the membrane and the concrete wall are lights that emit a glowing white light at night, further contributing to the sense that these buildings are as light as yurt tents. They seem to be able to move away at any time, responding to the nomadic nature of the grassland and minimizing the interference to the site environment. The two materials, the PTFE membrane and concrete, presenting a contrast of light and heavy, soft and hard, are joined at the curved undulating cornice formed by the inner courtyard. At the same time, the artificial interface composed of the turning and continuous walls and eaves has a strong sense of movement, making a dialogue with the vast natural grassland and the quiet site landscape. In the interior, the built-in furniture is designed in combination with arc walls and bay windows. Through the outer window and skylight, light penetrates into the room and frames the grassland and sky outside.

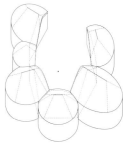

5 总平面草图 / Sketch of the site plan
6 总体模型（1/50）/ Model (1/50)
7 平面草图 / Plan sketch
8-9 构造草图 / Detail sketches
10 照明模型研究（1/30）/ Model study of lighting (1/30)

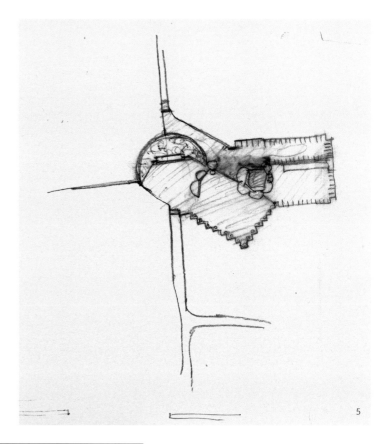

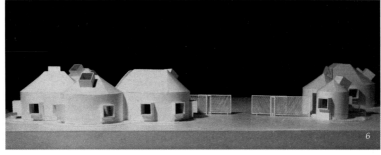

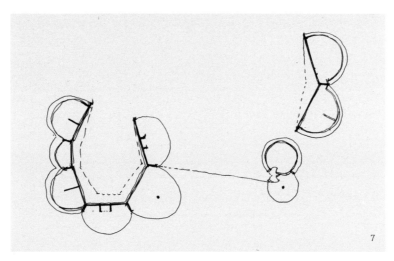

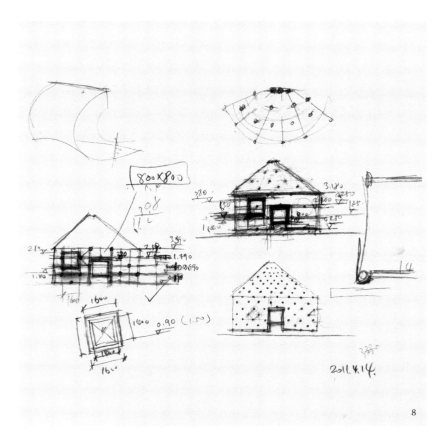

8

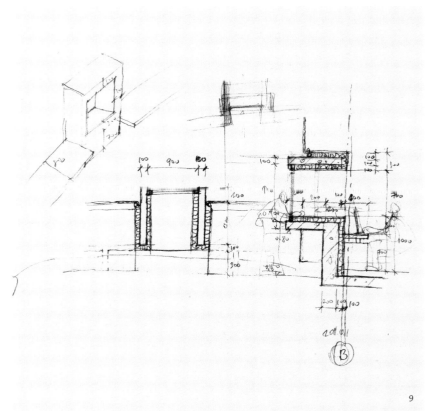

9

10

11

12

14 总平面图 / Site plan
15 南侧人视 / View from the south
16 南侧鸟瞰 / Aerial view from the south

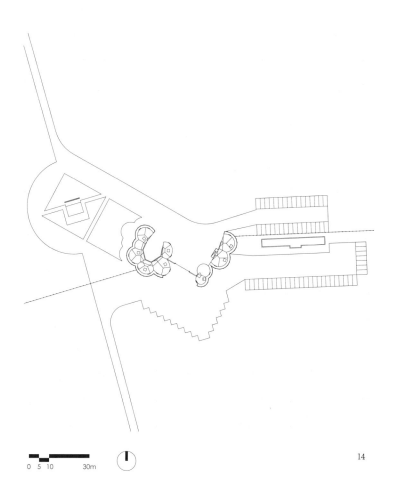

0 5 10   30m

14

15

16

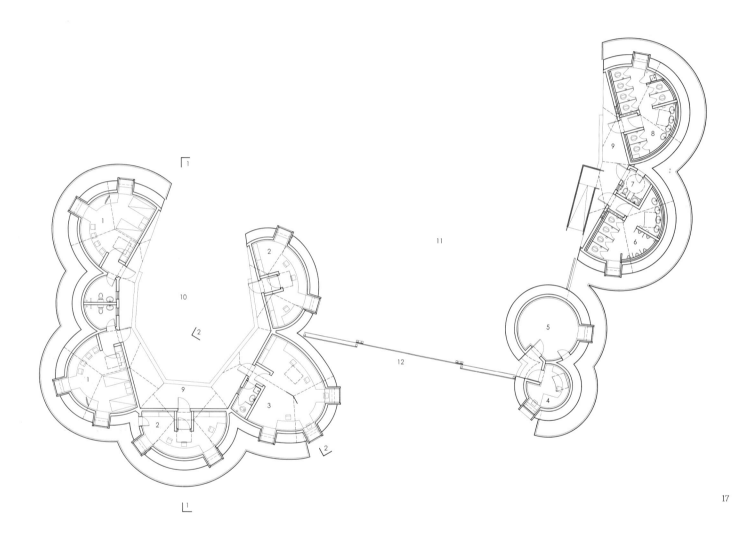

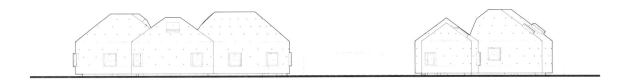

18

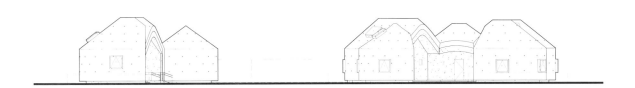

19

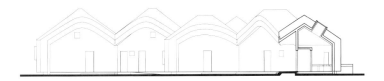

20

21

1 宿舍 / Dormitory
2 办公室 / Office
3 警卫监控室 / Security control room
4 售票处 / Ticket office
5 储藏间 / Storage
6 男卫生间 / Male lavatory
7 无障碍卫生间 / Barrier-free lavatory
8 女卫生间 / Female lavatory
9 檐廊 / Veranda
10 庭院 / Courtyard
11 广场 / Square
12 大门 / Gate

23

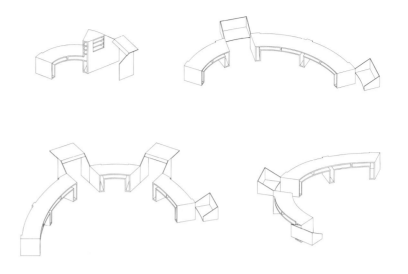

24

25

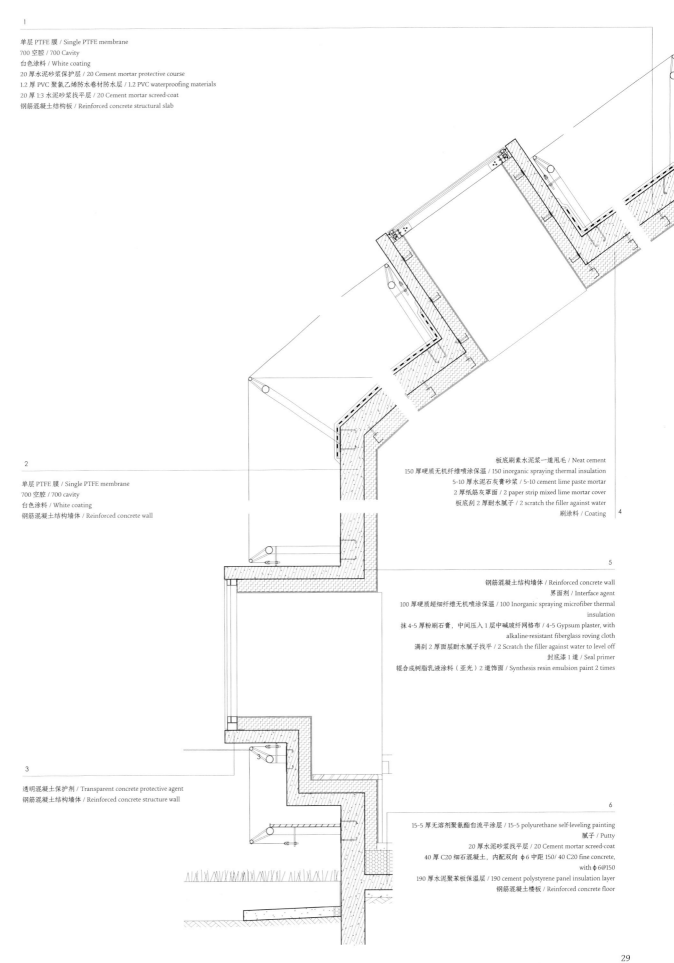

1
单层 PTFE 膜 / Single PTFE membrane
700 空腔 / 700 Cavity
白色涂料 / White coating
20 厚水泥砂浆保护层 / 20 Cement mortar protective course
1.2 厚 PVC 聚氯乙烯防水卷材防水层 / 1.2 PVC waterproofing materials
20 厚 1:3 水泥砂浆找平层 / 20 Cement mortar screed-coat
钢筋混凝土结构板 / Reinforced concrete structural slab

2
单层 PTFE 膜 / Single PTFE membrane
700 空腔 / 700 cavity
白色涂料 / White coating
钢筋混凝土结构墙体 / Reinforced concrete wall

3
透明混凝土保护剂 / Transparent concrete protective agent
钢筋混凝土结构墙体 / Reinforced concrete structure wall

板底刷素水泥浆一道甩毛 / Neat cement
150 厚硬质无机纤维喷涂保温 / 150 inorganic spraying thermal insulation
5-10 厚水泥石灰膏砂浆 / 5-10 cement lime paste mortar
2 厚纸筋灰罩面 / 2 paper strip mixed lime mortar cover
板底刮 2 厚耐水腻子 / 2 scratch the filler against water
刷涂料 / Coating   4

5
钢筋混凝土结构墙体 / Reinforced concrete wall
界面剂 / Interface agent
100 厚硬质超细纤维无机喷涂保温 / 100 Inorganic spraying microfiber thermal insulation
抹 4-5 厚粉刷石膏，中间压入 1 层中碱玻纤网格布 / 4-5 Gypsum plaster, with alkaline-resistant fiberglass roving cloth
满刮 2 厚面层耐水腻子找平 / 2 Scratch the filler against water to level off
封底漆 1 道 / Seal primer
辊合成树脂乳液涂料（亚光）2 道饰面 / Synthesis resin emulsion paint 2 times

6
15-5 厚无溶剂聚氨酯自流平涂层 / 15-5 polyurethane self-leveling painting
腻子 / Putty
20 厚水泥砂浆找平层 / 20 Cement mortar screed-coat
40 厚 C20 细石混凝土，内配双向 φ 6 中距 150/ 40 C20 fine concrete, with φ6@150
190 厚水泥聚苯板保温层 / 190 cement polystyrene panel insulation layer
钢筋混凝土楼板 / Reinforced concrete floor

29

2013

威海"Hiland·名座"
"HILAND·MINGZUO"
IN WEIHAI

　　威海"Hiland·名座"设计始于 2006 年 1 月，2013 年 11 月建成投入使用，位于山东半岛威海市城市干道海滨路和渔港路的交口处，邻近东部海滨，是一座以 SOHO 办公为主、兼具商业功能的房地产开发建筑。

　　根据当地的主导风向（夏季以东南风和南风为主、冬季以西北风为主），利用对流、气压差、热压"烟囱"效应等气流组织的基本原理，以低技的、简单直接的自然通风方式，在建筑内不同高度设定了多组下进上出、南进东出的"西南—东北"走势的"风径"，从而有效引入夏季风穿过建筑内部，以降温除湿，同时最大限度地回避冬季风对建筑的不利影响，并在设计过程中使用了 CFD 计算机模拟技术对风速、温度、湿度等进行舒适度模拟验证和校核，使建筑内尽量多的房间能够通过自然通风的方式降温，从而减少空调设备在夏季的使用。"风径"口部设置了可密闭的旋转玻璃门，根据外部天气的变化调整门的开闭情况以产生舒适的建筑小气候，实现建筑室内外空间的自然转换。"风径"的设置不仅实现了巧妙自然的建筑"环境调控"策略，有效地改善建筑内部小气候——降温除湿，也在建筑剖面上形成了位于不同高度的若干组内部中庭，服务于各自对应的 3—5 个自然楼层，成为积极的空中邻里交往空间和极佳的南向和东向对外观景界面和平台，使自然环境和人的活动在建筑中交融共存，也使建筑以一种开放的姿态和形象存在于城市之中。

　　建筑同时注重与城市环境的对话。红色坡屋顶的设计不仅呼应了威海市特有的红瓦坡顶建筑风格，同时也有效减少了建筑对基地南侧、东侧住宅的日照影响；立面块体图案的划分则依据原有周边的小尺度城市建筑，从而形成新旧建筑之间的对话和延续。为减少现场施工对密集的周边居民的影响，建筑外墙全部采用了预制混凝土复合保温挂板，这一措施同时保证了施工时效和质量，降低了建筑的自重，减少了建筑造价和外墙维护费用。

摄影 / Photographer:
张广源 / Zhang Guangyuan

1 概念草图 / Concept sketch
2 "风径" 模型（1/30）/ Model of the "wind tunnels" (1/30)
3 "风径" 图解 / Diagram of the "wind tunnels"

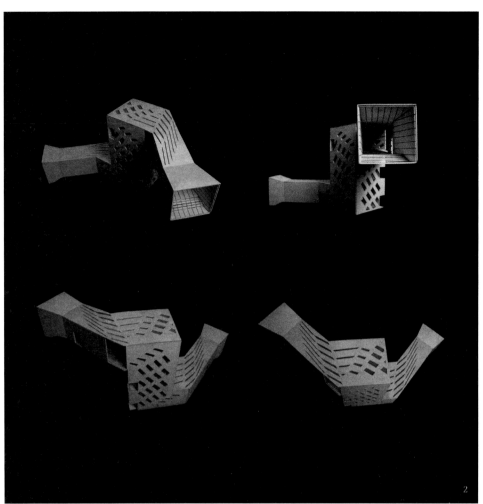

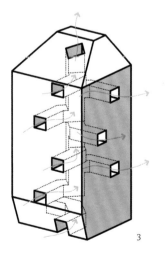

"Hiland · Mingzuo" in Weihai was initially designed in November 2009, completed and put into use in November 2013. Located at the intersection of Haibin Road and main city street of Yugang Road in the city of Weihai, on the Shandong Peninsula's eastern coast, it is a development primarily for SOHO units, with additional commercial space.

In order to take advantage of and adapt to the direction of changing local dominant wind patterns (mainly southeast wind and south wind in summer and northwest wind in winter), we considered the basic principles of air flow such as convection, pressure difference, and thermal pressure chimney effect to encourage natural building ventilation. We designed several groups of "wind tunnels" or ducts in the southwest-northeast direction (down in and up out, south in and east out) at different heights of the building. These tunnels provide a low-tech, simple, and energy efficient method of directing natural ventilation throughout the building using convection and differences in barometric pressure. In summer, these wind tunnels guide wind into the building for cooling and dehumidification. In winter, they minimize the effects of coastal wind to the building. Using Computational Fluid Dynamics (CFD) to simulate wind speed, temperature, and humidity during the design process to, verify and check the effect on climate inside the building. With this information, we maximized the number of rooms this system could be cooled by natural ventilation, decreasing the need to use air conditioning in summer. Depending on variations in outdoor weather, sealable revolving doors installed at the air inlets can be opened or closed to create a comfortable microclimate in the building. In addition to serving as a means to control the building's climate using cooling and dehumidification, the wind tunnels also form atriums of different heights throughout the building. These atriums can serve three or five corresponding floors. These open-air atriums encourage neighbors to gather and socialize. Platforms on the south and east atriums grant residents views of the sea. These areas invite an interaction between nature and human activity while offering an opening to the city.

The building also engages in a dialog with its surrounding urban environment. Its red pitched roof fits in with the style of red tiled pitched roofs in Weihai City. The roof of "Hiland · Mingzuo" can also reduce the influence of the sunlight routes to the neighbors. This new building's appropriate facade division also maintains continuity and dialogue with surrounding small buildings. The exterior walls are constructed with composite precast insulated concrete slabs and installed on-site. Therefore, the construction had minimal effect on the dense surrounding neighbors. These efficient construction methods reduced the building's intrusion on the neighborhood, cost, and ongoing maintenance cost.

5 "风径" 草图 / Sketch of the "wind tunnels"
6 平面草图 / Plan sketches
7 立、剖面草图 / Sketches of elevation and section
8-9 总体模型（1/100） / Models (1/100)

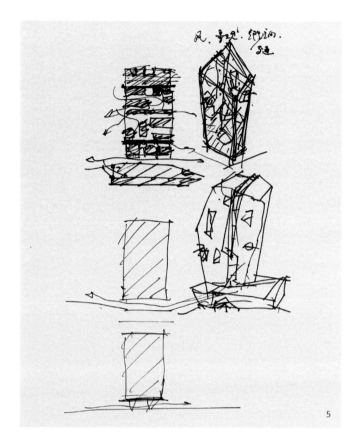

5

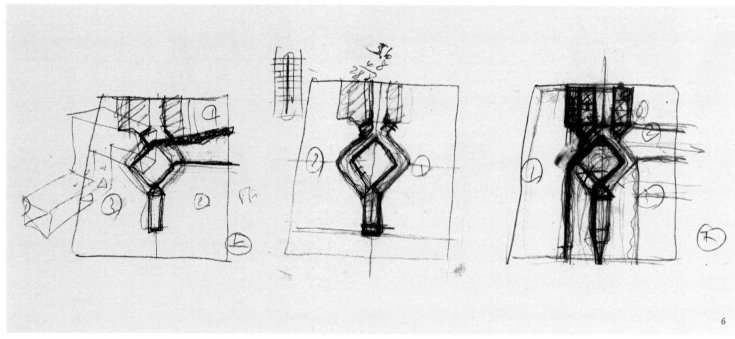

6

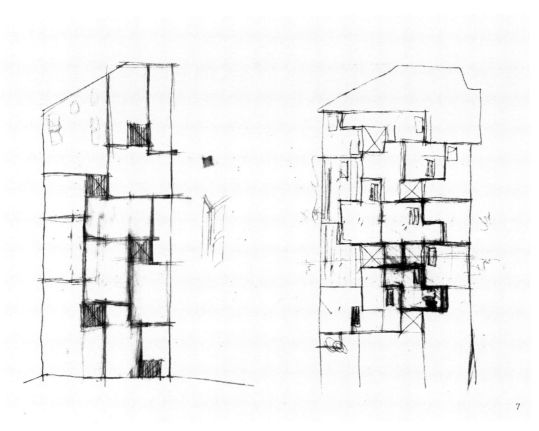

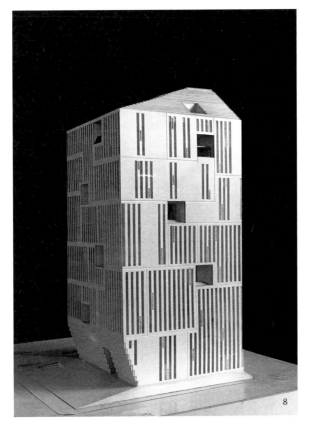

7

8

9

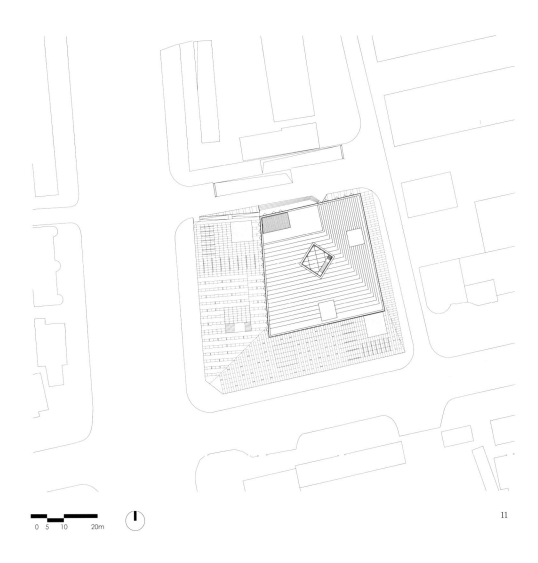

0 5   10   20m

11

0 2 5 10m

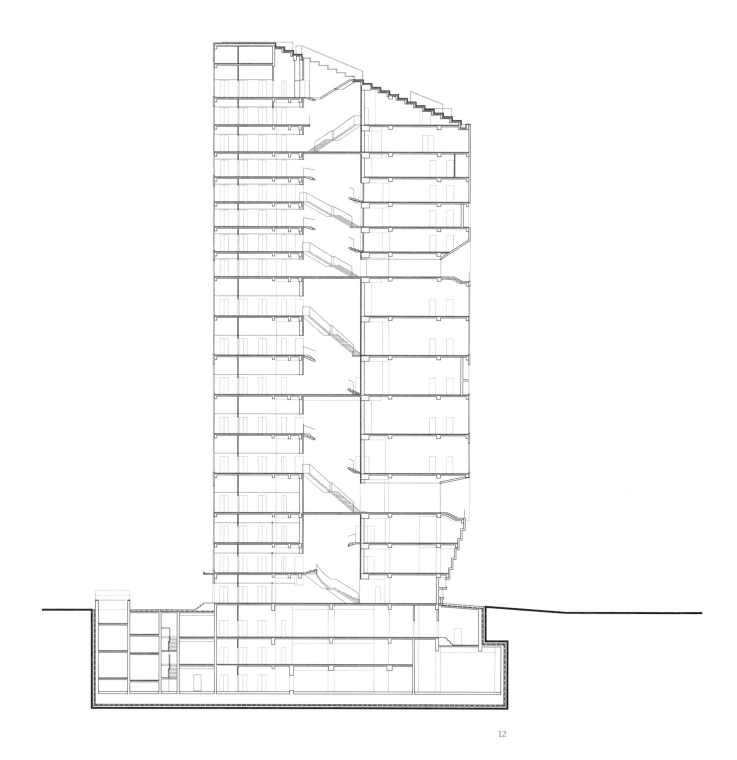

12

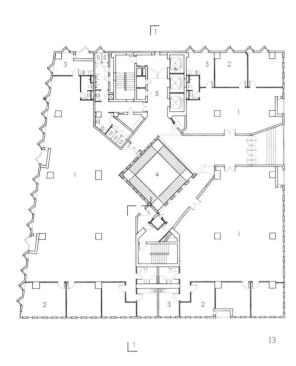

13

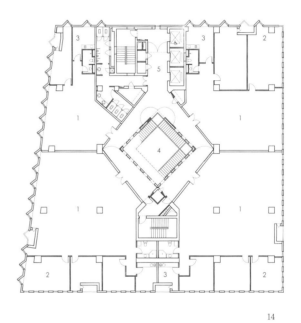

14

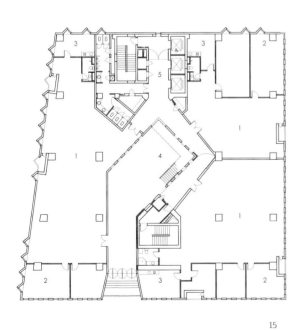

15

1 开敞办公区 / Open working area
2 办公室 / Office
3 服务间 / Service area
4 "风径" / "Wind tunnel"
5 电梯厅 / Lift hall

171

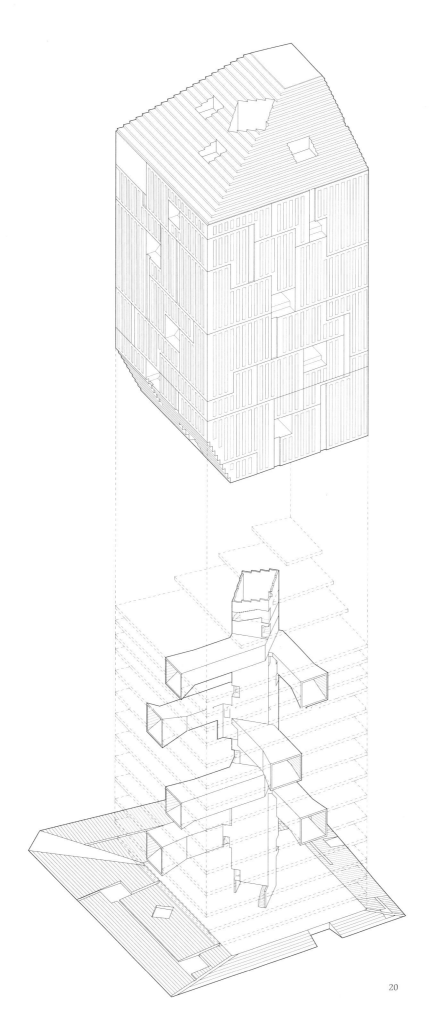

20

2013

绩溪博物馆
JIXI MUSEUM

　　绩溪博物馆设计始于 2009 年 11 月，2013 年 11 月建成投入使用，位于安徽绩溪县华阳镇旧城北部，基址曾为县衙，后建为县政府大院，因古城整体纳入保护修整规划，拆建为一座包含展示空间、4D 影院、观众服务、商铺、行政管理、库藏等功能区的中小型地方历史文化综合博物馆。博物馆施工过程中曾挖出县衙监狱部分的基础和排水沟等遗迹，设计修改将其实地保留为博物馆重要展览内容。这一项目尝试系统地呼应和解决旧城保护、更新与活化的问题，反映小城镇的文化姿态，铺陈和提升属于中国地方城市的文化历史品质与自信。 自开馆以来，博物馆作为公共空间已成为绩溪人日常生活中不可缺少的"城市客厅"，并吸引了来自全国各地的参观者，逐渐成为绩溪的城市名片。

　　绩溪得天独厚的山形水势、风土人文、村镇格局是博物馆的设计灵感之源。建筑设计基于对绩溪的地理环境、名称由来的考察和对徽派建筑与聚落的调查研究。整个建筑覆盖在一个由"屈曲并流，离而复合"的经线控制的连续屋面之下，结构剖面的组合变化使屋面轮廓此起彼伏，仿佛绩溪周边的山形脉络，是"绩溪之形"的充分演绎和展现。建筑不仅与周边民居乃至整个古镇自然地融为一体，也与周边山脉相互和应。

　　为尽可能保留用地内的 40 余株现状树木（特别是用地西北部一株 700 年树龄的古槐），建筑的整体布局中设置了多个庭院、天井和街巷，既营造出舒适宜人的室内外空间环境，也是徽派建筑空间布局的重释。建筑群落内利用庭院和街巷组织景观水系，东西两条水圳，汇于主入口庭院内的水面，有如绩溪地形中徽、乳两条水溪汇聚一体的形态。博物馆的室外空间对绩溪市民和游客开放，水圳汇流于前庭，成为入口游园观景空间的核心，由此开始，一条立体"观赏流线"将人们缓缓引导至建筑东南角的"观景台"，人们可以俯瞰建筑的屋面、庭院和秀美的远山。

　　成对排列、延伸的三角屋架单元，其坡度源自当地民居建筑，既营造出连续起伏的屋面形态，又暴露于室内，呈现出蜿蜒深远的内部空间。以徽州地区传统的"粉墙黛瓦"作为主要的建筑材料，并对其使用方式、部位和做法进行了大胆的转换，使之呈现出当代感。

摄影 / Photographer:
夏至、李哲、李兴钢、邱涧冰 / Xia Zhi, Li Zhe, Li Xinggang, Qiu Jianbing

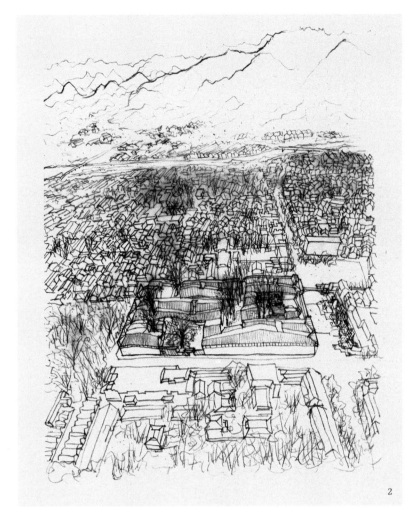

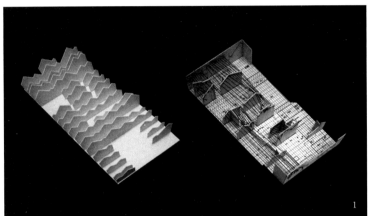

The Jixi Museum was initially designed in November 2009, completed and put into use in November 2013. Located in the north of the old city of Huayang Town, Jixi County, Anhui Province, the site was once the county's yamen (government), and later became the county government compound. Because the ancient town was integrated into the preservation and renovation plan, the compound was demolished and a medium-sized local history and culture museum was built on the site with exhibition space, 4D cinema, visitors' service, shops, administrative offices, storage and other functional areas. During the museum's construction process, the foundation, drainage ditch and other ruins of the county prison were excavated. The museum's excavated site was memorialized as an exhibition through the followed design modifications. This project reflects possible solutions to reconcile the interests of preservation of historical cities with urban renewal and stimulus. A goal for this project is to reflect the proud cultural heritage of Jixi while reviving its cultural and historical standing. Since its opening, the museum has become a public space and an indispensable "urban living room" in the daily life of Jixi people. The Jixi Museum attracts visitors from all over the country and has become the city's cultural calling card.

Jixi's unique geography, which includes mountains, rivers, people and culture along with its historic village plan inspired the museum's design. We investigated the origins of the region's namesake, huipai (Anhui) style of architecture, its ancient settlements, and its relationship with Jixi's geography. The building is covered by a continuous roof rolling contours that "bends and flows, splits and merges". The combination of structural sections forms the undulating roof outline, like the mountain ridge around Jixi, which is the full embodiment of Jixi's topography. The building is not only naturally integrated with the surrounding residential buildings and the whole ancient town, but it also interacts with the surrounding mountains.

Courtyards, patios, and paths are carefully arranged in the museum's layout to retain and incorporate more than 40 existing trees on the site, including a 700-year old locust tree in the northwest. We not only create pleasant indoor and outdoor spaces, but also reinterprets the spatial layout of huipai buildings. The building, its courtyards, and paths blend with the complex's water elements. The east and west water channels join together in the main entrance courtyard, like the two converging streams in Jixi County. The outdoor public space of the museum is open to Jixi citizens and tourists. The water channel converges in the vestibule and becomes the core of the viewing space of the entrance garden. Here, sightline slowly guides people to a platform at the southeast corner of the building, on which people may overlook the roof, courtyards, and beautiful distant mountains. The triangular roof truss units were arranged and extended in pairs with sloped angles derived from local residential buildings. The roof design not only creates an undulating roof on the exterior, but also forms a spacious, deep, and winding interior underneath. The traditional white walls and black tiles in Huizhou area have been used as the main building materials. However, the bold transformations of the materials in usage, position and practice make the building present a contemporary sense.

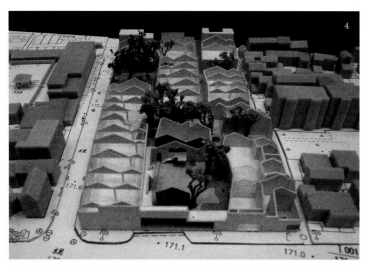

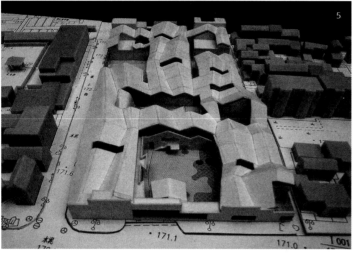

4-5 结构及屋顶模型（1/300）/ Models of the structure and the roof (1/300)
6 场地模型（1/500）/ Site model (1/500)
7 总体模型（1/200）/ Model (1/200)

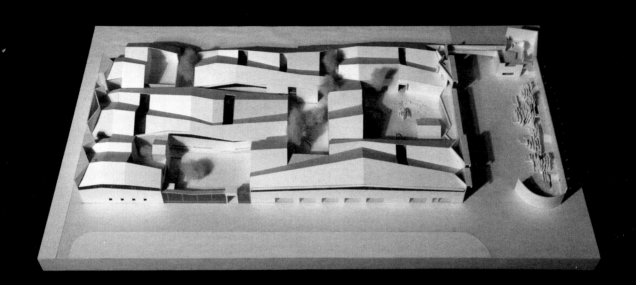

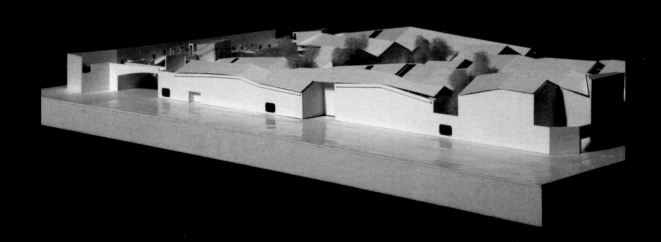

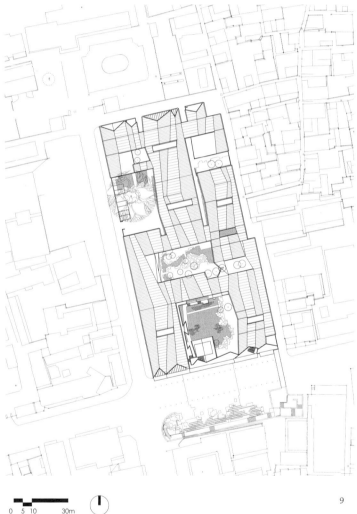

9

11 屋面、庭院与远山 / View of the roof, the courtyards and distant mountains
12 一层平面 / The 1st floor plan
13 剖面 1-1/ Section 1-1

0 2 5 10m

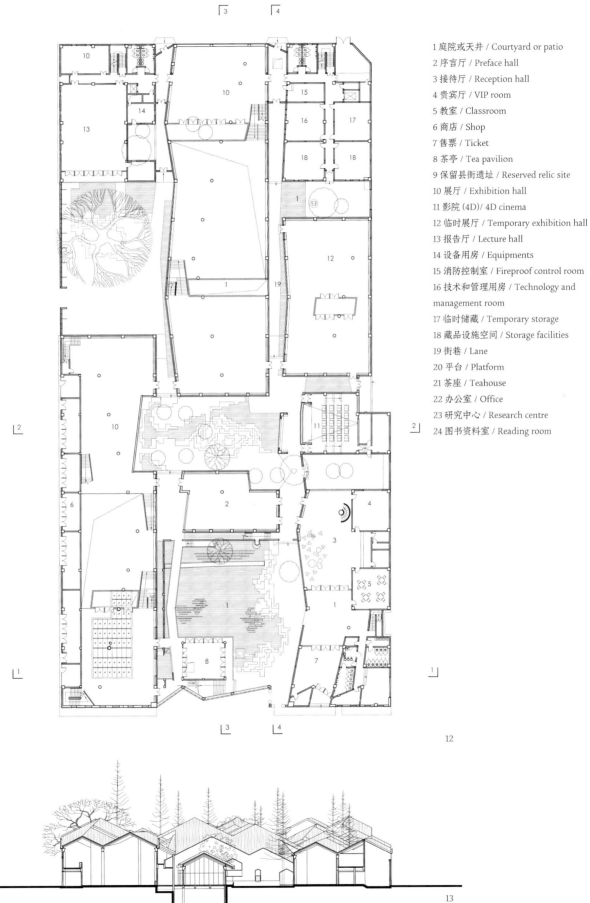

1 庭院或天井 / Courtyard or patio
2 序言厅 / Preface hall
3 接待厅 / Reception hall
4 贵宾厅 / VIP room
5 教室 / Classroom
6 商店 / Shop
7 售票 / Ticket
8 茶亭 / Tea pavilion
9 保留县街遗址 / Reserved relic site
10 展厅 / Exhibition hall
11 影院 (4D)/ 4D cinema
12 临时展厅 / Temporary exhibition hall
13 报告厅 / Lecture hall
14 设备用房 / Equipments
15 消防控制室 / Fireproof control room
16 技术和管理用房 / Technology and
management room
17 临时储藏 / Temporary storage
18 藏品设施空间 / Storage facilities
19 街巷 / Lane
20 平台 / Platform
21 茶座 / Teahouse
22 办公室 / Office
23 研究中心 / Research centre
24 图书资料室 / Reading room

12

13

16

17

18 "山"院 / View of the tree courtyard
19 二层平面 / The 2nd floor plan
20 剖面 2-2/ Section 2-2

19

20

0   5   10   20m

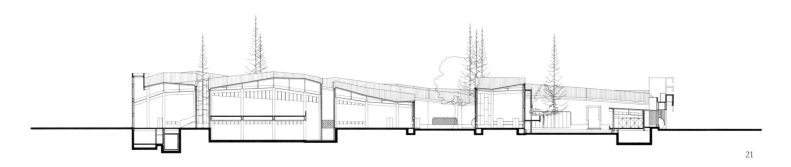

21

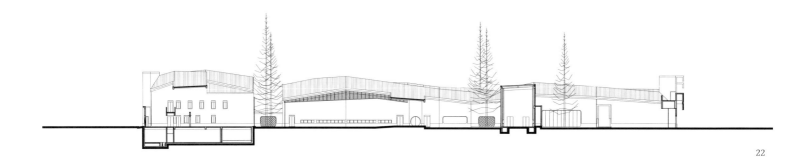

22

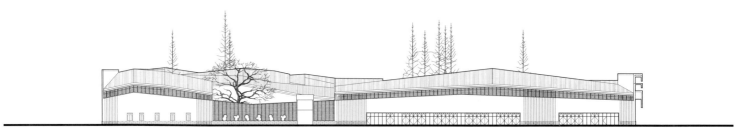

23

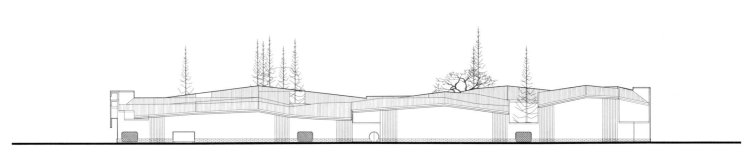

24

28

30 施工中的屋顶 / The roof under construction
31 墙身及屋顶详图 / Details of the wall and the roof
32 施工中的展厅 / The exhibition hall under construction
33 施工现场鸟瞰 / Aerial view of the construction site

30

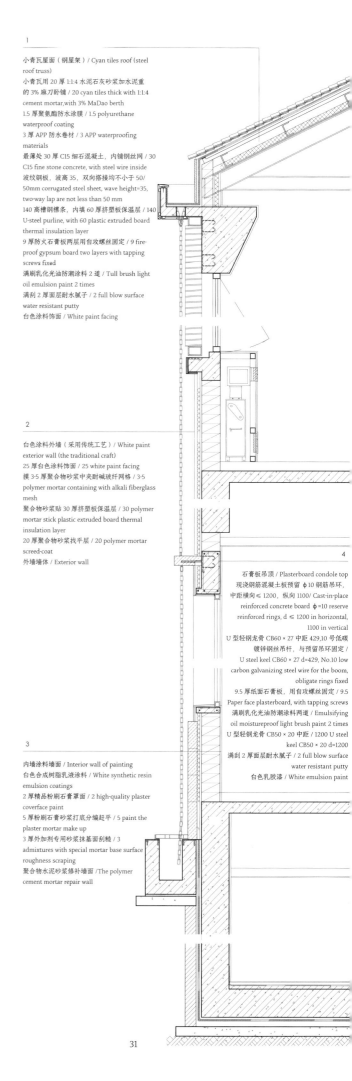

1

小青瓦屋面（钢屋架）/ Cyan tiles roof (steel roof truss)
小青瓦用 20 厚 1:1:4 水泥石灰砂浆加水泥重的 3% 麻刀卧铺 / 20 cyan tiles thick with 1:1:4 cement mortar,with 3% MaDao berth
1.5 厚聚氨酯防水涂膜 / 1.5 polyurethane waterproof coating
3 厚 APP 防水卷材 / 3 APP waterproofing materials
最薄处 30 厚 C15 细石混凝土，内铺钢丝网 / 30 C15 fine stone concrete, with steel wire inside
波纹钢板，波高 35，双向搭接均不小于 50/ 50mm corrugated steel sheet, wave height=35, two-way lap are not less than 50 mm
140 高槽钢檩条，内填 60 厚挤塑板保温层 / 140 U-steel purline, with 60 plastic extruded board thermal insulation layer
9 厚防火石膏板两层用自攻螺丝固定 / 9 fire-proof gypsum board two layers with tapping screws fixed
满刷乳化光油防潮涂料 2 道 / Tull brush light oil emulsion paint 2 times
满刮 2 厚面层耐水腻子 / 2 full blow surface water resistant putty
白色涂料饰面 / White paint facing

2

白色涂料外墙（采用传统工艺）/ White paint exterior wall (the traditional craft)
25 厚白色涂料饰面 / 25 white paint facing
摸 3-5 厚聚合物砂浆中夹耐碱玻纤网格 / 3-5 polymer mortar containing with alkali fiberglass mesh
聚合物砂浆贴 30 厚挤塑板保温层 / 30 polymer mortar stick plastic extruded board thermal insulation layer
20 厚聚合物砂浆找平层 / 20 polymer mortar screed-coat
外墙墙体 / Exterior wall

3

内墙涂料墙面 / Interior wall of painting
白色合成树脂乳液涂料 / White synthetic resin emulsion coatings
2 厚精品粉刷石膏罩面 / 2 high-quality plaster coverface paint
5 厚粉刷石膏砂浆打底分编赶平 / 5 paint the plaster mortar make up
3 厚外加剂专用砂浆抹基面刮糙 / 3 admixtures with special mortar base surface roughness scraping
聚合物水泥砂浆修补墙面 /The polymer cement mortar repair wall

4

石膏板吊顶 / Plasterboard condole top
现浇钢筋混凝土板预留 φ10 钢筋吊环，中距横向 ≤ 1200，纵向 1100/ Cast-in-place reinforced concrete board φ =10 reserve reinforced rings, d ≤ 1200 in horizontal, 1100 in vertical
U 型轻钢龙骨 CB60 × 27 中距 429,10 号低碳镀锌钢丝吊杆，与预留吊环固定 / U steel keel CB60 × 27 d=429, No.10 low carbon galvanizing steel wire for the boom, obligate rings fixed
9.5 厚纸面石膏板，用自攻螺丝固定 / 9.5 Paper face plasterboard, with tapping screws
满刷乳化光油防潮涂料两道 / Emulsifying oil moistureproof light brush paint 2 times
U 型轻钢龙骨 CB50 × 20 中距 / 1200 U steel keel CB50 × 20 d=1200
满刮 2 厚面层耐水腻子 / 2 full blow surface water resistant putty
白色乳胶漆 / White emulsion paint

31

196

2015

鸟巢文化中心
BIRD'S NEST
CULTURAL CENTER

鸟巢文化中心设计始于 2012 年 10 月，2015 年 3 月建成投入使用，位于国家体育场"鸟巢"北部，是一个在保护奥运遗产的前提下，按照预定的赛后经营计划对原有体育场局部空间进行改造的项目，旨在依托国家体育场的建筑特色，构建一个文化艺术交流平台。由外部的下沉庭院及入口引道和内部的零层及地下一层多功能空间组成，可举办展览、会议、表演、宴会等多种类型和规模的活动，已成为北京重要的公共文化、艺术、体育、公益主题空间，是奥运遗产保护利用和预留奥运后运营空间改造的成功范例。

"鸟巢"的整体设计中存在一个对应外部钢结构的不规则轴网和对应内部混凝土结构的放射状轴网，新的设计引入了一个黄金分割比矩形格网体系，叠加在原有"鸟巢"的结构主导轴网之上，并将此格网进一步扩展为模数控制下的矩形板块系统，同时作用于平面和立面，以此为基础，建立起一套新的语汇系统，将室内外空间元素（墙、地、顶）一体化处理，并保留和强化原建筑极具表现力的结构构件，生成与"鸟巢"形象空间相协调又并置和凸显自身特征的室内空间和景观环境。

在下沉庭院起造抽象的山景水景，竖向层叠的"片岩"假山（内有阶梯蹬道、平台、景亭）和水平拼合的水面、池岸、浮桥、平台、亭榭均由模数控制的清水混凝土单元板块堆叠成形，与爬藤、花草、树木相结合，营造出兼具古意和当代感的山水园林，可踱步、攀爬、登临、驻留、凝望。在同样的模数控制下，水平片状单元继续向零层大厅室内曼延，形成连接零层及地下一层标高的叠落状混凝土台地，兼具展示和观演功能，并在上部形成沿"鸟巢"基座（即大厅顶板）标高由外向内逐渐抬升的直角折线放射状木制板块单元吊顶。"台地大厅"是一个层高达 9.5 米的通高空间，外侧是沿放射状轴网布置的垂直柱林，内侧是一列由地上延伸入地下空间的"鸟巢"主次钢结构柱，呈现不规则的尺寸和斜向，犹如巨大的钢制雕塑装置。透过斜钢柱体形成的巨型网格，视线可回望聚焦于眼前一片叠落而下的混凝土台地"山坡"和远处大厅外下沉庭院的"片石"池岸与假山，意境深远。

摄影 / Photographer:
孙海霆、李兴钢 / Sun Haiting, Li Xinggang

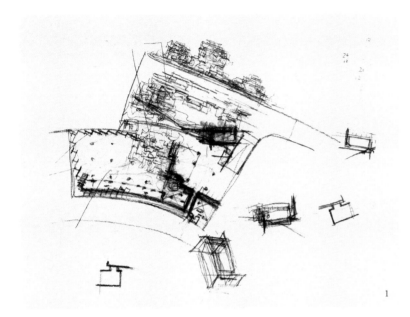

The Bird's Nest Cultural Center, which was initially designed in May 2014 and put into use in March 2015, is located in the north of the National Stadium—Bird's Nest. This is a renovation project of part of the original stadium space in accordance with the scheduled post-Olympic games operation plan. It aims to build a cultural and artistic exchange platform based on the architectural characteristics of the National Stadium. Consisting of an external sunken courtyard, an entrance road and an interior multi-functional space below the ground floor, it can hold various types and scales of exhibitions, conferences, performances, and public benefit activities. Now, the Bird's Nest Cultural Center has become an important multi-use public space for culture, art, sports, and banquets in Beijing. It is a successful example of a structure that preserves its Olympic heritage, while, after renovation, can continue to be an operational space after the games.

The Bird's Nest maintains an irregular grid of external steel structures and a radial grid of corresponding internal concrete structures. The new design introduces a golden ratio rectangular grid system, superimposed on the original "Bird's Nest" structure. The grid is further extended to the rectangular plate system to support both the plane and facade. The Cultural Center creates its own new architectural vocabulary that strengthens the original buildings, while its interior and exterior spatial elements integrate and harmonize with the Bird's Nest.

In the sunken courtyard, an abstract view of water and mountains is created. The vertically stacked "schist" rock formations (with steps, platforms and pavilions inside) and the horizontal water surface, pool bank, pontoon bridge, platform, and pavilion are stacked by modular pale concrete plates. The project integrates climbing vines, flowers, and trees to create a landscape garden with both ancient and contemporary influences. Visitors can cross, climb, board, stay, and observe on the courtyard. Under the same modular control, the horizontal unit continues to spread into the interior of the ground floor hall, forming a stacking concrete platform connecting the ground floor and the basement level, which has both display and exhibition functions. In the upper floors, a rectangular creased radiating wooden plate ceiling is formed, which rises gradually from the exterior to the interior along the elevation of the Bird's Nest base (the hall roof). The "Terminal Hall" is a 9.5-meter tall space. The exterior is a vertical column forest arranged along a radial grid. The interior are the stadium's primary and secondary steel structural columns extend from the above-ground into the underground, showing irregular dimensions and diagonals, and like huge steel sculptures. Through the giant grid formed by the inclined steel cylinders, the sight line focuses on the "hill slope" of the concrete platform in front of the collapsed hills, the "grain stone" pool, and rock formations in the exterior courtyard to instill a sense of profundity.

4 "片岩"假山草图 / Sketch of the "schist" rock formations
5 室内草图 /Sketch of the interior space
6-7 平面草图 / Plan sketches

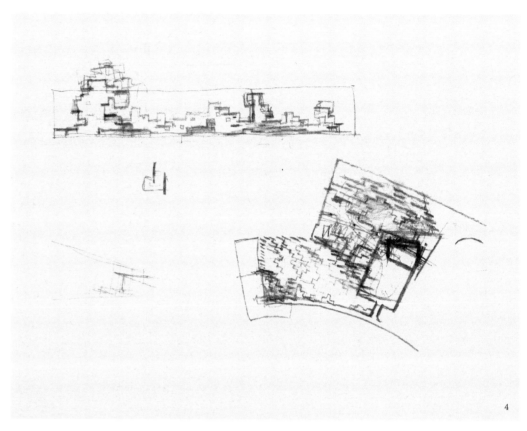

4

5

6

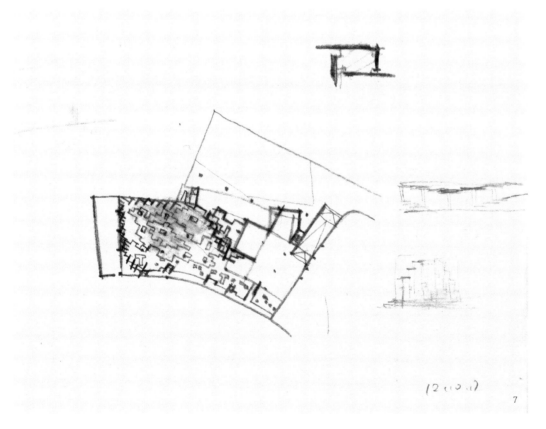

(2 つ))

7

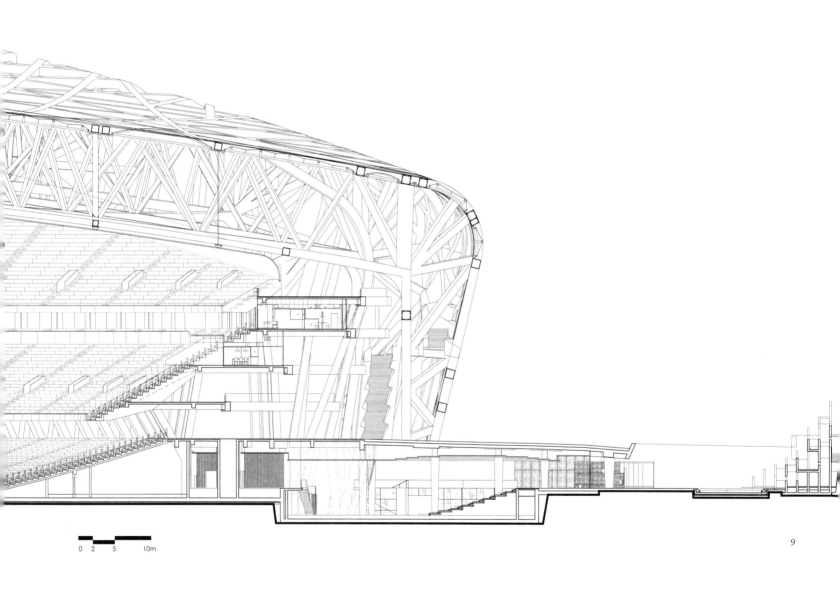

9

0 2 5 10m

10

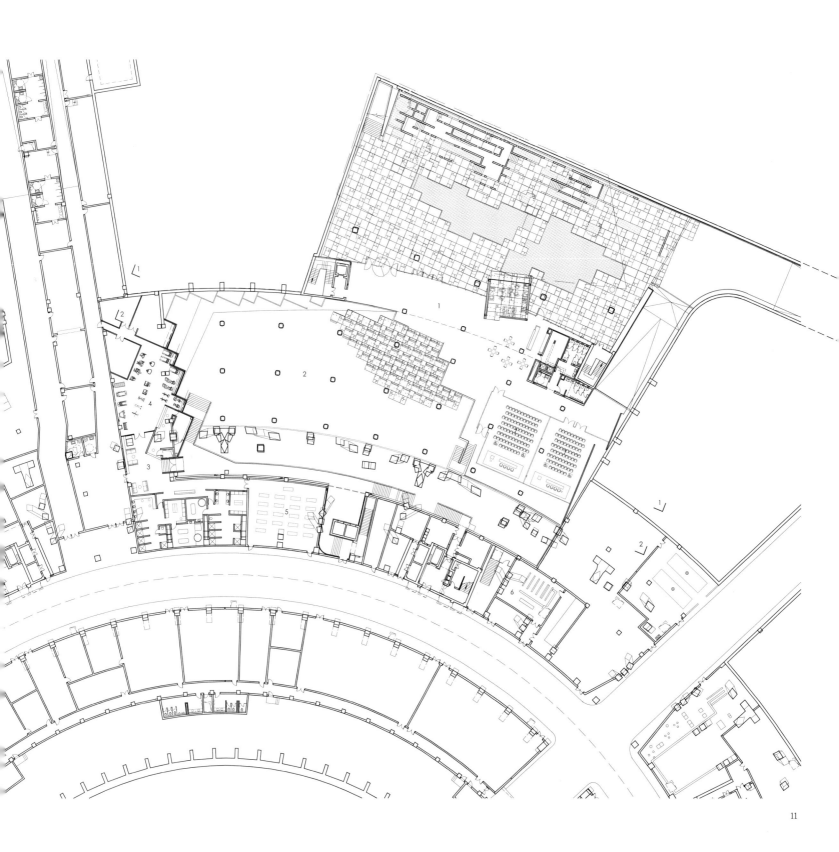

11

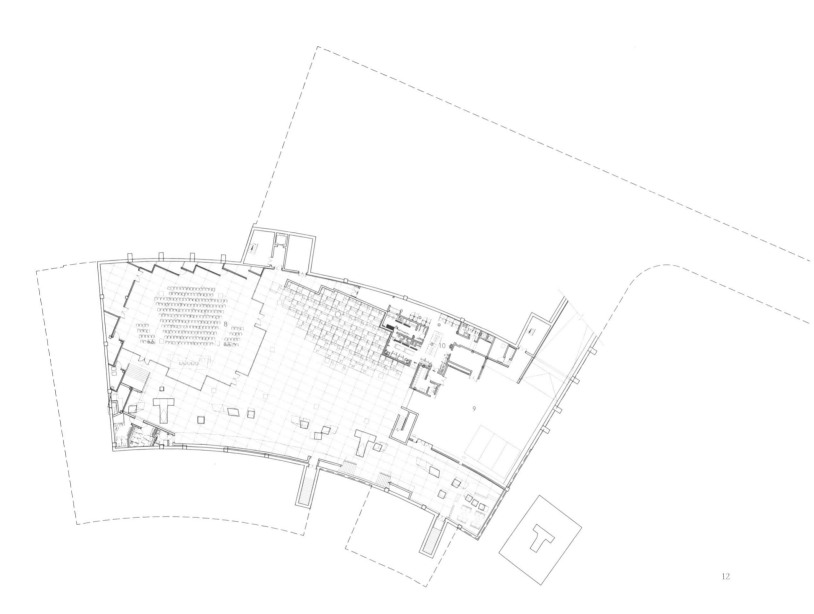

12

1 入口门厅 / Entrance hall
2 多功能厅 / Multi-functional hall
3 健身房门厅 / Entrance hall of the gym
4 健身房 / Gym
5 体操房 / Gymnastics room
6 机房 / Equipment room
7 临时会议室 / Meeting room
8 临时报告厅 /Lecture hall
9 车库 / Garage
10 厨房 / Kitchen

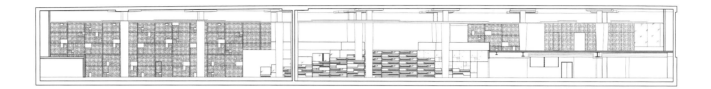

13

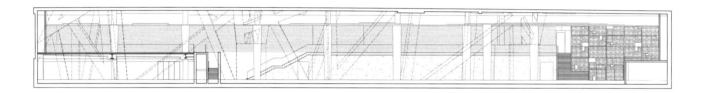

14

15

18 "片岩" 假山平面 / Plan of the "schist" rock formations
19 "片岩" 假山剖面 1-1/ Section 1-1 of the "schist" rock formations
20 "片岩" 假山剖面 2-2/ Section 2-2 of the "schist" rock formations
21 "片岩" 假山人视 / View of the "schist" rock formations

0   2   5   10m

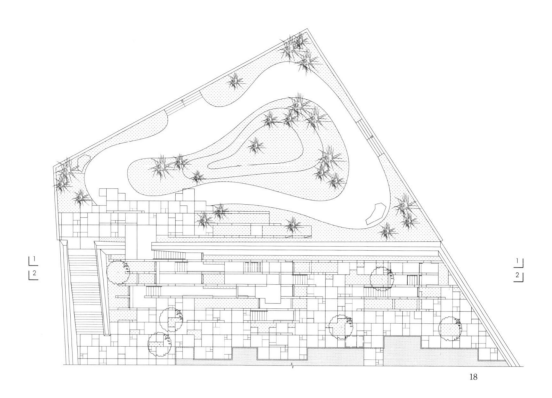

18

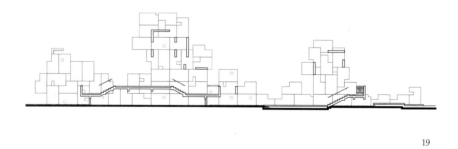

19

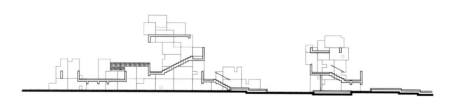

20

2015

商丘博物馆
SHANGQIU MUSEUM

　　商丘博物馆设计始于 2008 年 12 月，2015 年 5 月竣工，位于商丘西南城市新区，收藏、陈列和展示商丘的历代文物、城市沿革和中国商文化历史。商丘位于海河平原和淮河平原之间的黄河冲积平原，地势低平，长久以来形成了典型的"居高筑台、城墙护堤、蓄水坑塘"的洪涝适应性景观和城镇形态。博物馆的整体布局和空间序列是对以商丘归德古城为代表的黄泛平原古城池典型形制和特征的呼应和再现，博物馆犹如一座微缩的古城。建筑上下叠层喻示商丘"城压城"的古城考古埋层结构，也体现自下而上、由古至今的陈列布局。

　　博物馆主体由三层体量叠加的展厅组成，周围环以水面和庭院，水面和庭院之外是层层叠落的台地绿植和其外围高起的堤台（下面设室外展廊），文物、业务和办公用房组成 L 形体量，设置于西北角堤台之上。设南北东西四门，主入口设在南门，其他三门各有贵宾、临展、办公等功能空间。

　　参观者由面向阏伯路的大台阶和坡道登临堤台，沿南面凌水引桥而由中部序言厅入"城"，自下而上，沿中央十字大厅（喻城中十字主街）中的坡道陆续参观各个展厅，最后到达屋顶平台，可由建筑主体各角眺望台，与不同方向的著名古迹——阏伯台、归德古城、隋唐大运河码头遗址等遥遥相望，怀古思今。周围下沉式的景观台地是对文物发掘现场的模拟，使得建筑主体犹如是被发掘出来一般。古象形文"商"字的含义是"高台上的子姓族人"，博物馆所形成的层层高起的堤岸、平台和其上的参观者组合成为"高台上的子姓族人"意象，再现"商"字的古老渊源，也将历史和当下联为一体。一系列精心组织的建筑和景观元素，形成了在上下、内外、近远之间往复变化的空间叙事，使参观者游观尽致，完成对这个微缩之城及其所承载延伸的古城历史的完整体验。

　　博物馆大量采用了一种廉价的"鲁灰"石材，作为建筑内外空间的主要界面材料，但受到博物馆汉画像石藏品的启发，每块石材均做了磨切外边 + 中间烧毛的处理，错缝拼挂，使细节显得考究。在室内加入了树脂实木面板材，与石材采用统一规格，增强了室内空间的温暖和舒适感。

摄影 / Photographer:
夏至 / Xia Zhi

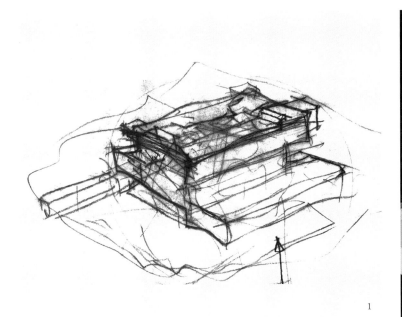

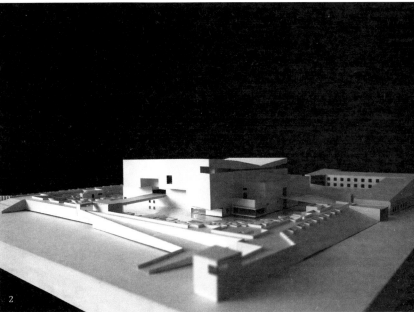

1

2

The Shangqiu Museum, initially designed in December 2008 and completed in May 2015, is located in the new district in the southwest of the city and was built for the collection, display, and exhibition of local historical relics. The museum also traces the Shangs' urban evolution and its cultural history. Shangqiu is located in the low and flat alluvial plain of the Yellow River between the Haihe plain and the Huaihe plain. For a long time, it has formed a typical flood adaptive landscape and urban engineering that places the city on a plateau, surrounded by a city wall on an embankment, with ponds for storing water. The overall layout and the spatial sequence of the museum echoes and represents the typical forms and characteristics of ancient Guide city, a representative of ancient city of Huangfan. The museum is itself akin to a miniature ancient city. With floors "stacked" atop each other, the museum suggests an excavation site with the tiers of a buried architectural site. The museum's layout and galleries showcase the temporal transition from the ground up, beginning with its ancient history to ascending to its present.

The main body of the museum consists of three stacked exhibition halls surrounded by water and courtyards. The landscaped courtyards have outdoor terraces, lined with exhibition corridors, above which are the peripheral platforms where visitors can peer over. The L-shaped buildings in the northwest corner of the platform are house relics and administrative offices. Four gates are set at four different directions. The main entrance is located at the south gate with the other three gates used as VIP area, temporary exhibitions entrance, staff entry, and for other purposes.

After climbing the platform through major steps and ramps facing E'bo Road and walking through the south bridge above river, visitors enter the "city" from the preface hall. Now at an "intersection," visitors can walk to exhibition halls in different directions along ramps. When arriving at the roof platform, visitors can overlook the well-known monuments such as E'botai platform, Guide ancient city, the site of the Sui and Tang Dynasties Grand Canal Pier in different gazebos at each corner of the building, recalling history and rethinking the present situation. The surrounding sunken, landscaped tiers simulate the excavation site of cultural relics. The ancient pictograph 商 (Shang) means "members of the clan with a family name of Zi standing on a high platform", which is symbolized by visitors standing on the tiers of levee and platforms. This reinterprets the origin of the character and elegantly combines the past and the present. A series of elaborately organized elements of architecture and landscaping form a sequence of narrative spaces high and low, inside and outside, and near and far. These alternating spatial changes enable visitors to fully experience the history of Shangqiu by circulating through its museum.

Inspired by its collection of Han Dynasty images of Shandong grey stone, we used the abundant and affordable stones as the primary exterior and interior cladding. Each stone slab was treated with surface honing, singed at the center, and installed in an elegant brick pattern. The resin-coated, solid wood panels with unified specifications, same as the stones, were adopted to enhance a sense of warmth and comfort in the indoor spaces.

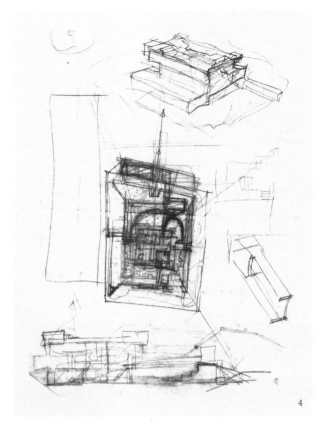

4

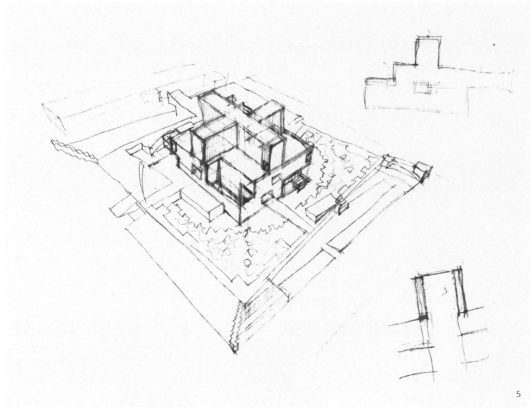

5

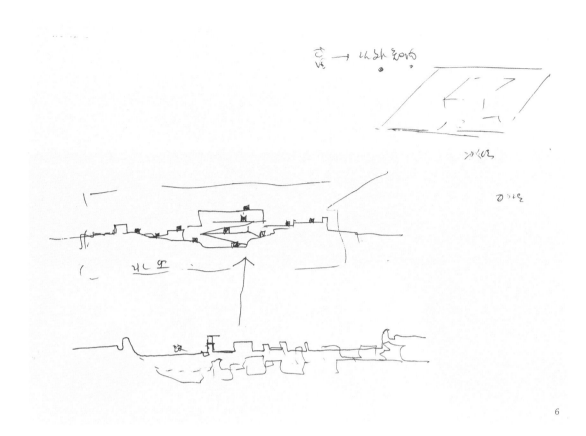

6

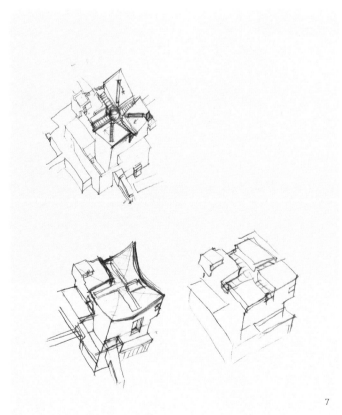

7

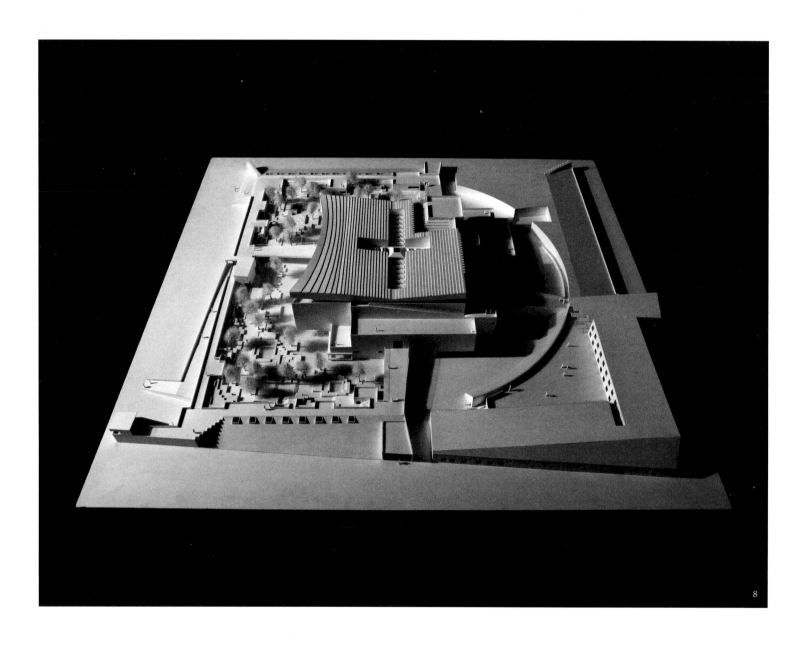

8 总体模型（1/250）/ Model (1/250)
9 剖面模型（1/100）/ Section model (1/100)

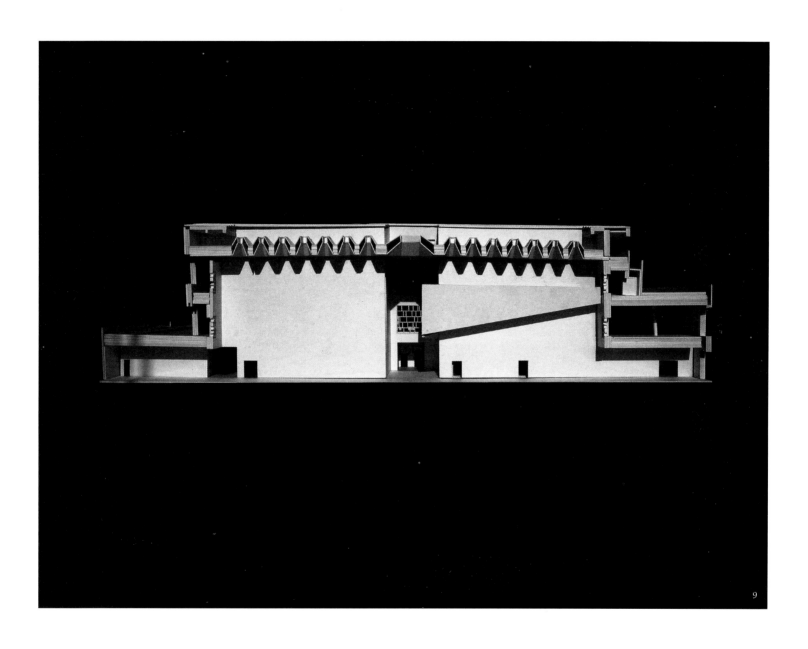

10

0 10 20   50m

11

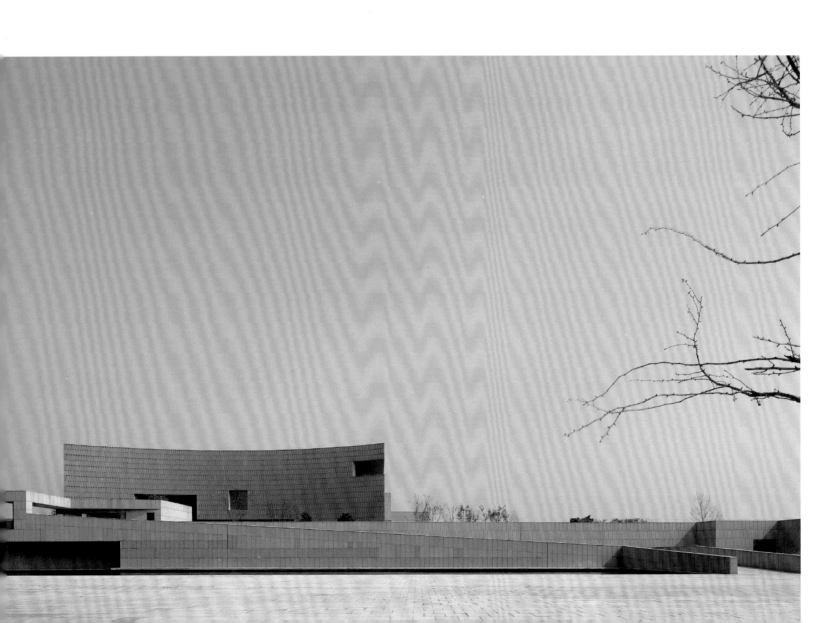

13

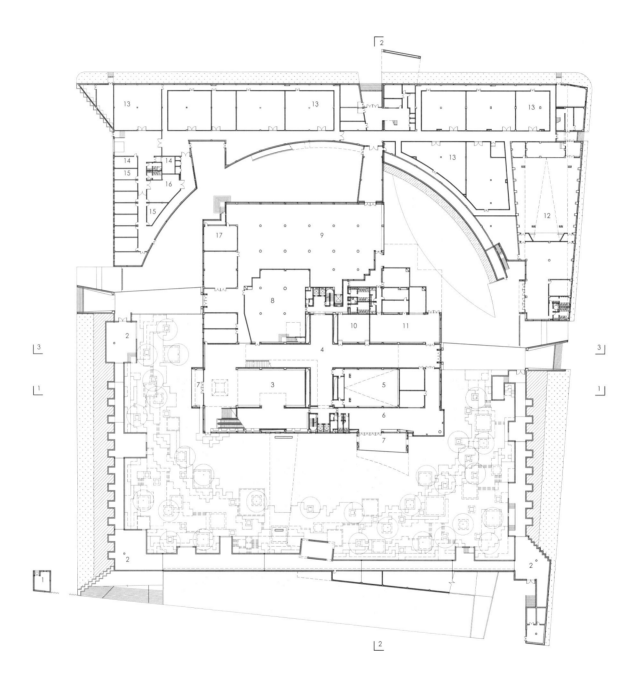

1 警卫室 / Guard house

2 室外展廊 / Outdoor gallery

3 序言厅 / Preface hall

4 共享大厅 / Main hall

5 放映厅 / Cinema

6 茶餐厅 / Tea restaurant

7 亲水平台 / Waterside terrace

8 展厅 / Exhibition hall

9 多功能大厅 / Multi-functional hall

14

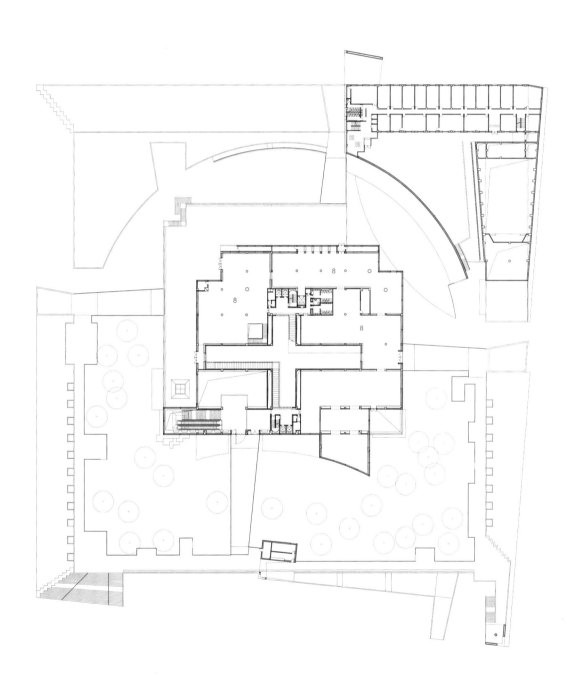

15

0 5 10   20m

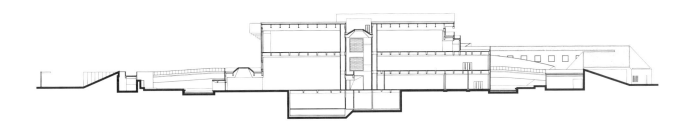

19

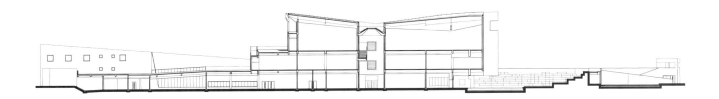

20

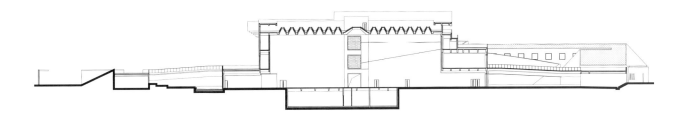

21

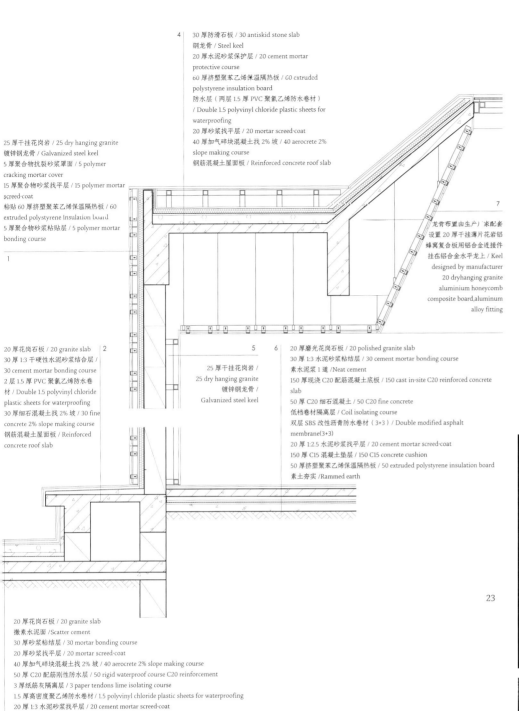

4 | 30 厚防滑石板 / 30 antiskid stone slab
钢龙骨 / Steel keel
20 厚水泥砂浆保护层 / 20 cement mortar protective course
60 厚挤塑聚苯乙烯保温隔热板 / 60 extruded polystyrene insulation board
防水层（两层 1.5 厚 PVC 聚氯乙烯防水卷材） / Double 1.5 polyvinyl chloride plastic sheets for waterproofing
20 厚砂浆平层 / 20 mortar screed-coat
40 厚加气碎块混凝土找 2% 坡 / 40 aerocrete 2% slope making course
钢筋混凝土屋面板 / Reinforced concrete roof slab

25 厚干挂花岗岩 / 25 dry hanging granite
镀锌钢龙骨 / Galvanized steel keel
5 厚聚合物抗裂砂浆罩面 / 5 polymer cracking mortar cover
15 厚聚合物砂浆找平层 / 15 polymer mortar screed-coat
粘贴 60 厚挤塑聚苯乙烯保温隔热板 / 60 extruded polystyrene Insulation board
5 厚聚合物砂浆粘贴层 / 5 polymer mortar bonding course

1

7 | 龙骨布置由生产厂家配套设置 20 厚干挂薄片花岗岩蜂窝复合板用铝合金连接件挂在铝合金水平龙上 / Keel designed by manufacturer 20 dryhanging granite aluminium honeycomb composite board,aluminum alloy fitting

20 厚花岗石板 / 20 granite slab
30 厚 1:3 干硬性水泥砂浆结合层 / 30 cement mortar bonding course
2 层 1.5 厚 PVC 聚氯乙烯防水卷材 / Double 1.5 polyvinyl chloride plastic sheets for waterproofing
30 厚细石混凝土找 2% 坡 / 30 fine concrete 2% slope making course
钢筋混凝土屋面板 / Reinforced concrete roof slab

5 6

25 厚干挂花岗岩 / 25 dry hanging granite
镀锌钢龙骨 / Galvanized steel keel

20 厚磨光花岗石板 / 20 polished granite slab
30 厚 1:3 水泥砂浆粘结层 / 30 cement mortar bonding course
素水泥浆 1 道 / Neat cement
150 厚现浇 C20 配筋混凝土底板 / 150 cast in-site C20 reinforced concrete slab
50 厚 C20 细石混凝土 / 50 C20 fine concrete
低档卷材隔离层 / Coil isolating course
双层 SBS 改性沥青防水卷材（3+3） / Double modified asphalt membrane(3+3)
20 厚 1:2.5 水泥砂浆找平层 / 20 cement mortar screed-coat
150 厚 C15 混凝土垫层 / 150 C15 concrete cushion
50 厚挤塑聚苯乙烯保温隔热板 / 50 extruded polystyrene insulation board
素土夯实 /Rammed earth

20 厚花岗石板 / 20 granite slab
撒素水泥面 /Scatter cement
30 厚砂浆粘结层 / 30 mortar bonding course
20 厚砂浆找平层 / 20 mortar screed-coat
40 厚加气碎块混凝土找 1% 坡 / 40 aerocrete 2% slope making course
50 厚 C20 配筋刚性防水层 / 50 rigid waterproof course C20 reinforcement
3 厚纸筋灰隔离层 / 3 paper tendons lime isolating course
1.5 厚高密度聚乙烯防水卷材 / 1.5 polyvinyl chloride plastic sheets for waterproofing
20 厚 1:3 水泥砂浆找平层 / 20 cement mortar screed-coat
150 厚配筋 C20 现浇混凝土上板 / 150 cast in-site C20 reinforced concrete slab
100 厚 C15 混凝土垫层 / 100 C15 concrete cushion
素土夯实 / Rammed earth

23

24

2015

唐山"第三空间"
THE "THIRD SPACE"
IN TANGSHAN

　　唐山"第三空间"综合体设计始于 2009 年 2 月，2015 年 6 月竣工，位于唐山市建设北路东侧，其用地东侧紧邻一片平行排列的工人住宅。建筑的朝向、布局和塔楼及裙房的体量、形状几乎完全由日照计算得出，以满足严格的法规要求。两栋平行的百米高楼顺着西南阳光的入射方向旋转了一个角度，朝向东南，裙房的屋顶也被"阳光通道"切成了锯齿形，其东侧留出一个带状的空地花园。

　　唐山是一座历史悠久的冀东之城，也是中国近现代工业的摇篮。1976 年的罕见大地震将整个城市几乎夷为平地，震后重建的新城带有强烈的快速、简单和人工化的特征。与唐山大量的住宅板楼构建出的"平行城市"面貌不同，"第三空间"呈现为一个向高空延伸的立体城市聚落——76 套在城市中垂直叠摞的"别业"宅园。置身于具有平均主义色彩的震后住宅中间，"第三空间"试图在当代的保障性住宅和郊区别墅之外寻找另外的可能性，填补颇具现实意义的居住类型上的空白，并暗示了一种自发性建造的可能性；在城市集合住宅普遍平庸的环境中，探索了居住单元空间的个性化特征。不同于通常住宅中的标准层，楼板以错层的方式层叠交错，形成连续抬升的地面标高，在区分出公共与私密的不同使用功能的同时，创造出丰富的空间层次，赋予了居住生活多样的体验性和视觉化特征，在城市中心区的高密度环境中实现"理想居住"。立面上形态、大小、朝向各异的亭台小屋不仅收纳了城市风景，而且向城市展示出了一个个生动的生活舞台，自身成为城市中的新景观。

　　"第三空间"的主体结构是钢筋混凝土框架剪力墙结构，由复式单元挑出于立面之外的大小不一的亭台，是安装在混凝土密肋板上的一个个独立的轻钢结构。主体混凝土结构为现浇建造，亭台钢结构则为工厂预制、现场装配吊装，外立面墙体则主要采用了预制 GRC 面材及轻钢龙骨干挂式结构。这样一种混凝土与钢结构、现制与预制的"混合建造"方式取决于当地的施工技术水平和"第三空间"的特定空间与结构特征。悬挑亭台及北向大阳台内部表面采用的各色不规则瓷片，既强化了"城市聚落"的特征，也是对本地著名的陶瓷历史与产业的呼应和致敬。

摄影 / Photographer:
张广源、孙鹏、李兴钢 / Zhang Guangyuan, Sun Peng, Li Xinggang

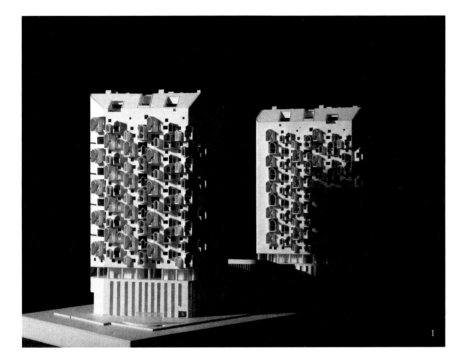

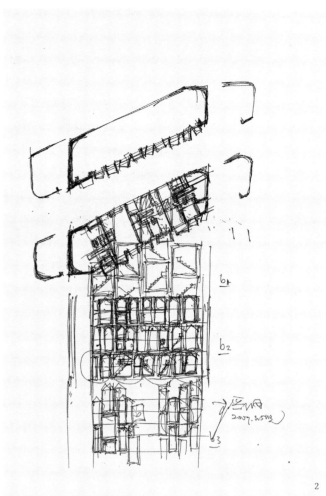

The "Third Space" in Tangshan, which was initially designed in February 2009 and completed in June 2015, is located on the east side of North Jianshe Road, Tangshan City. To the site's east are several rows of workers' houses aligned north to south in parallel. The complex's direction, layout, the tower's volume and the podium's shape are nearly all determined by the calculation of sunlight in order to fulfill the rigorous sunlight standard. We rotated the two parallel, 100-meter tall slab-type buildings along a south-east orientation to accommodate the direction of the incoming southwest sunlight. The roof of the podium was cut into a zigzag shape. A ribbon-shaped garden is created in the east side.

As a city with a long history in the east of Hebei Province, Tangshan is regarded as the cradle of modern Chinese industry. In 1976, a rare earthquake almost leveled the whole city, and the new city was quickly rebuilt after the earthquake. Although buildings were well-built, the speed of the design and construction failed to provide the new cityscape with much character. The "Third Space" is a three-dimensional city settlement with 76 vertical stacked villas extending to the sky. In the middle of the uniform, post-earthquake houses, the "Third Space" tries to find another possibility in addition to the existing affordable housing and suburban single-family homes. Filling this residential gap, the "Third Space" shows that housing can be unique, utilitarian, and built quickly. Amidst the monotonous apartments surrounding it, the "Third Space" explored the possibility that high-rise living could be personalized. Distinguished from Tangshan's repetitive apartment slabs arranged in parallel on their sites, this project is characterized by staggered floors forming continuous ascension of height level in each unit. Each unit, despite having separate public and private functions, creates rich spatial levels that visually and experientially combine all the aspects of idealized living in a dense city center. All the duplex apartment units are stacked vertically, each with outdoor balconies of different sizes and orientations. These units become frames through which residents overlook the city-scape. Serving as lively stages of life open to the city, they become symbols of a dense, vertically-arranged urban neighborhood and a new urban landscape.

The "Third Space" is structured with reinforced concrete frames and shear walls. The pavilions cantilevered on the facade are independent, light-weight steel structures installed on the concrete ribbed slabs. The pavilions are prefabricated and assembled on site. The facade is mainly made of prefabricated Glass Fiber Reinforced Concrete (GRC) panels hanging on steel frames. Such a hybrid construction plan of mixing concrete and steel structures with pre-made and prefabricated approaches was dependent on local construction technologies and influenced the project's specific spatial and structural characteristics. The irregular porcelain pieces of various colors used on the internal surface of the cantilevered pavilions and the north-facing balconies not only highlight the characteristics of urban settlement, but also respond to and honor the city's local ceramic heritage and industry.

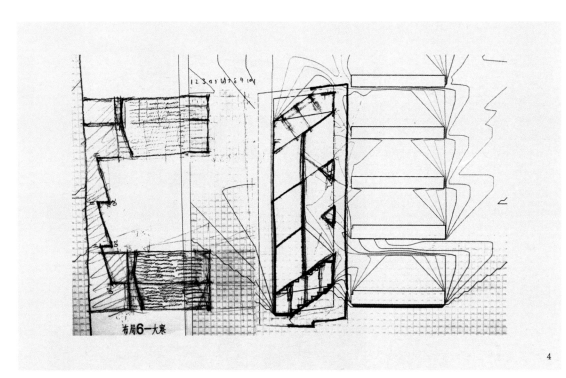

布局6—大寒

4

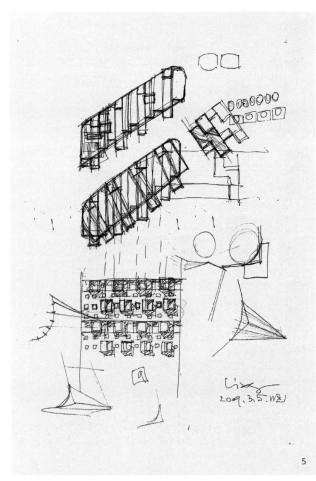

Lix
2009.3.5.W刻

5

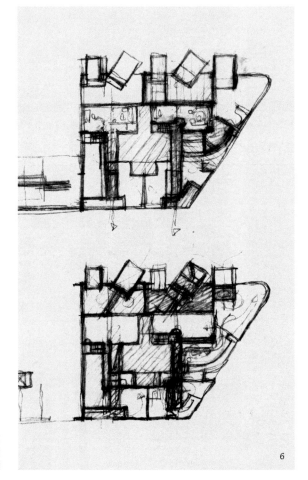

6

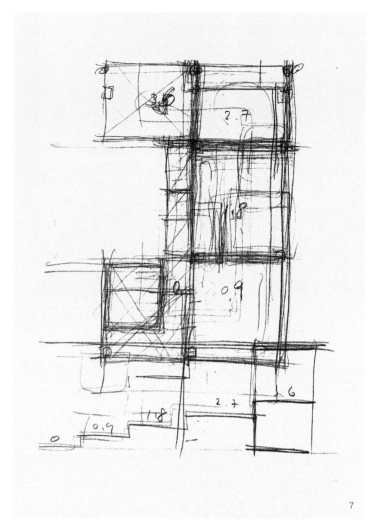

7

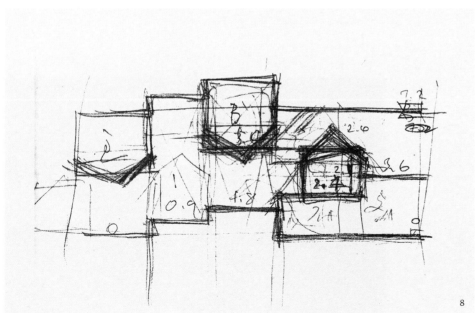

8

9

10

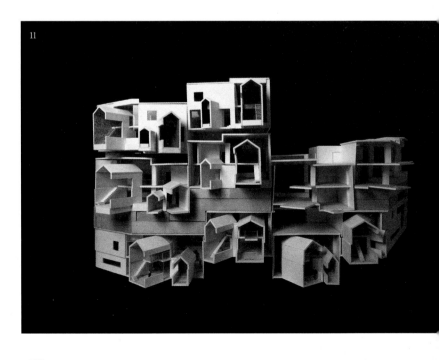

11

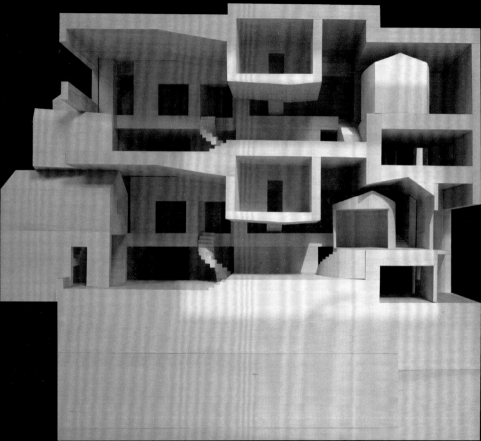

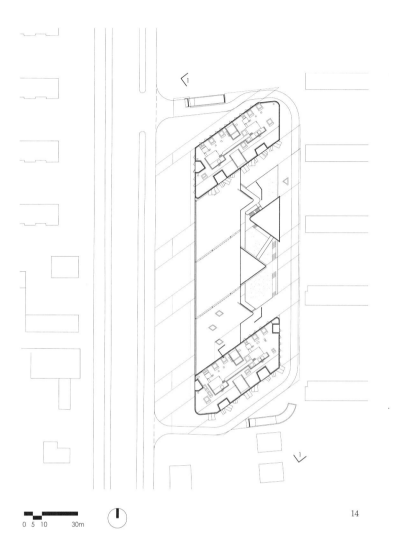

0 5 10   30m

14

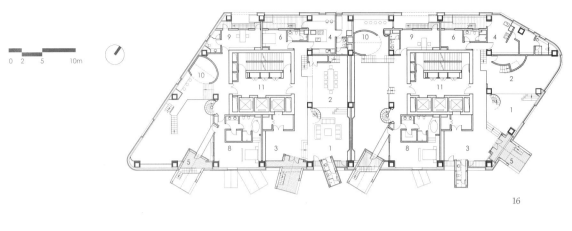

0 2 5 10m

16

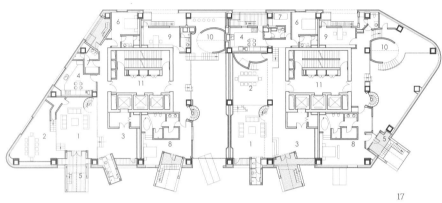

17

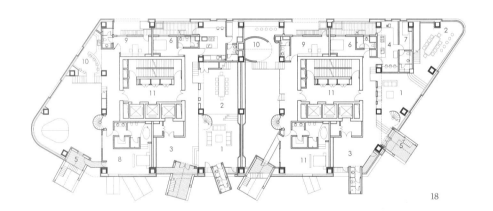

18

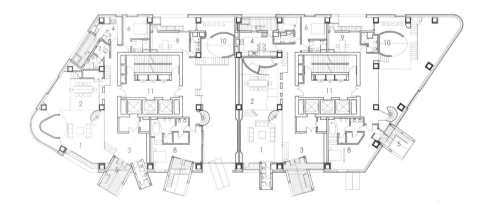

19

1 客厅 / Living room

2 餐厅 / Dining room

3 门厅 / Entrance

4 厨房 / Kitchen

5 景观亭台 / Sightseeing pavilion

6 客卧 / Guest room

7 佣人房 /Servant room

8 主卧 / Master room

9 书房 / Study room

10 红酒收藏 / Wine collection & bar

11 电梯厅 / Lift hall

20

21

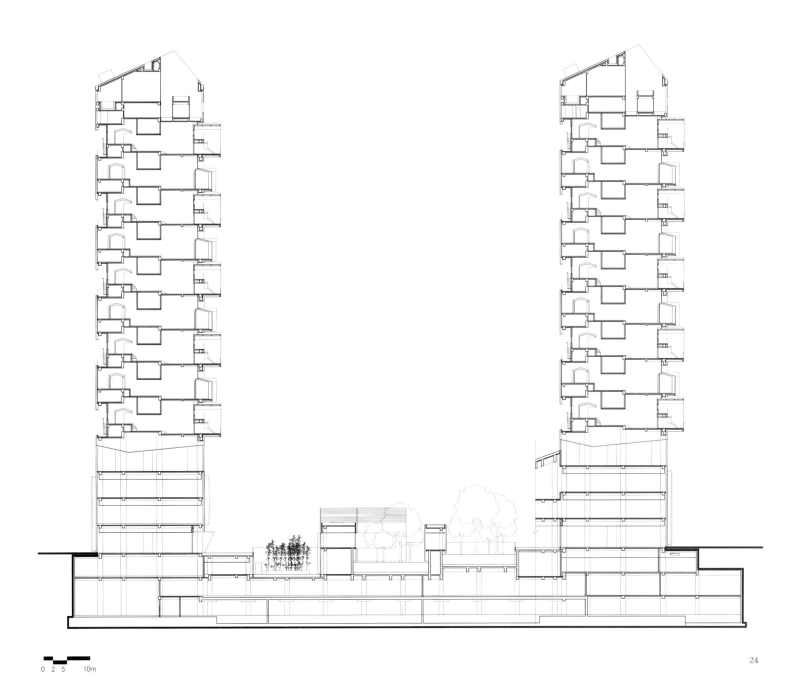

25

26

0 1 2   5m

27

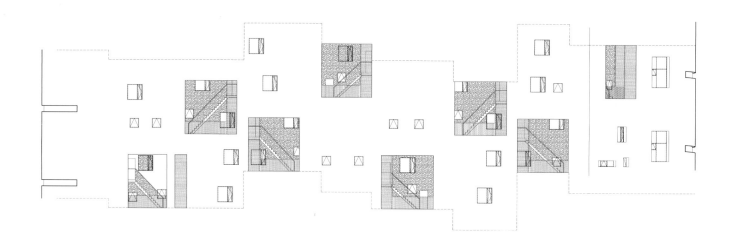

28

2015

国家体育场 2015 年世锦赛注册中心
REGISTRATION CENTER OF IAAF WORLD CHAMPIONSHIPS, BEIJING 2015

国家体育场 2015 年世锦赛注册中心设计始于 2015 年 3 月, 2015 年 8 月建成投入使用, 位于国家体育场——"鸟巢"用地内的室外热身场北侧, 承担第 15 届世界田径锦标赛组委会、志愿者、媒体及转播商等人员的注册制证任务, 并设置部分配套办公用房, 赛后按商业办公功能预留。在施工工期异常紧张的特殊情况下, 选用适宜的结构体系和材料以及预制装配的施工方法, 以控制建设周期和保证建筑品质, 并如期完成交付使用。

建筑尽量平展并占满不规则而狭促的基地, 以降低高度、弱化建筑体量, 北低南高, 中部微微隆起, 东西两面的檐口压低, 整个建筑低调地匍匐在国家体育场北部, 以保持原有的国家体育场的周边视野和环境氛围, 同时为所在场地赋予新的特质。为了能同时适应赛时的大量人流以及赛后的商业运营, 室内为大空间, 中间隆起部分形成变化净高的室内空间, 为未来设置局部夹层预留了条件, 设置高侧窗满足中间部分的采光要求, 南部屋面突起的"亭子"形成朝向国家体育场的景观视野, 东西两侧周边高起的台基与挑出压低的屋面檐口共同形成了一圈近人尺度的漫步檐廊。

建筑结构以 8 米的钢结构柱网为基本尺度, 局部变化以适应建筑轮廓的变化, 屋顶采用 2.6 米的正交网格体系减小屋顶的跨度, 下部结构柱并不直接接触屋面, 而是通过"树状"的斜向支撑连接屋顶网格体系的交点与柱子, 来承担屋面的重量, 每根柱子上的斜撑分布在两个不同标高处, 形成伞柱结构, 顺应屋顶标高变化, 自然形成了各组斜撑间的标高起伏, 斜撑因此是 8 米柱网和 2.6 米网格间的自然过渡, 斜撑的角度与长度的变化与柱子的秩序形成了对比。玻璃幕墙之外的部分屋顶改变了结构构件的方向, 形成连续低平的檐下空间。

场地北部的绿地内现存几棵大树, 并放置着国家体育场基座上的那种景观条石, 于是在建筑屋顶留出了洞口让树木主干可以穿过, 条石则作为公共休息座椅的同时也起到了界定入口空间的作用。

摄影 / Photographer:
孙海霆 / Sun Haiting

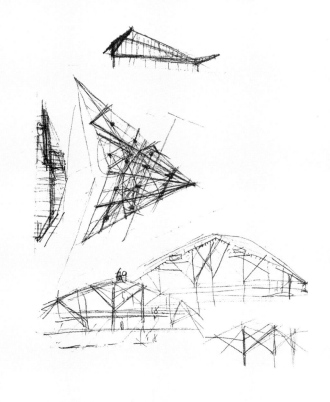

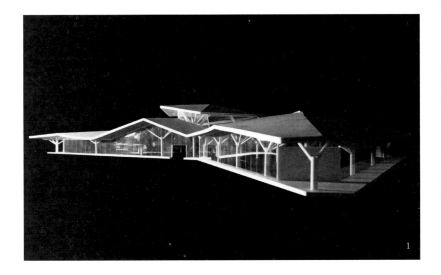

The Registration Center of 2015 IAAF World Championships in Beijing was initially designed in March 2015 and put into operation in August 2015. The Registration Center is located to the north of the outdoor warm-up field at the National Stadium–the Bird's Nest. The center was build as an area for registration and credential certification of organizing committee members, volunteers, the media, broadcasters, and administrative office workers. Because of the international importance of the site and tight construction schedule, we needed to select appropriate structural systems and materials, and prefabricated construction methods to control the construction cycle, ensure building quality, and complete construction on schedule.

The building is designed to be as flat as possible as it occupies an irregular and narrow site. In an attempt to reduce its height and decrease its volume, the building is lower in the north and higher in the south, slightly uplifted in the middle, with low cornices on the eastern and western sides. The building settles to the north of the National Stadium with a modest profile to draw focus of one's peripheral vision towards the stunning originality of the National Stadium. Nonetheless, the Registration Center contributes new design elements to the site. In order to accommodate to a large number of visitors during the competition and the commercial uses after competition, the interior is a large space with a raised, center portion with variable heights that includes vertical space for construction of a mezzanine tier in the future. High side windows illuminate the center of the building. The raised "pavilion" on the south roof offers views of the National Stadium. A raised platform around the east and west sides is enclosed by a low corniced roof to form a portico ideal for walking.

The building is framed by an 8-meter steel column network that was modified locally to adapt to the contours of the building. The 2.6-meter orthogonal grid system was used to reduce the span of the roof. The structural columns do not support the roof directly, but pass through tree-like diagonal supports that intersect with the roof's grid with the pillars to bear the weight of the roof. The diagonal braces on the pillars are distributed at two different heights, forming the umbrella-column structure. Following the changing slopes of the roof, the elevation of the diagonal braces fluctuates naturally. The diagonal braces play a transitional role between the 8-meter column grid and the 2.6-meter grid. The changing angle and length of the diagonal braces contrast with the order of columns. The direction of structural elements changes at the portion of the roof outside the glass curtain wall, creating a continuous walking space under the eaves.

Several tall trees occupy the green space in the north of the site. A landscaping stone bench from the base of the National Stadium is also placed in the green space. The building roof has an opening where tree trunks are able to protrude. Stone seating at the entrance also redefines the lobby.

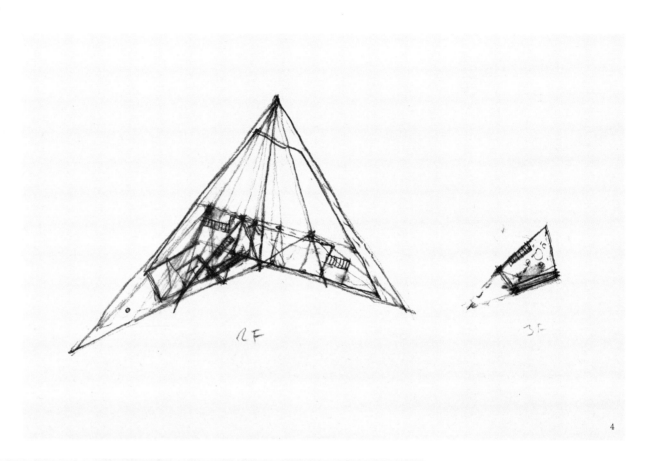

4

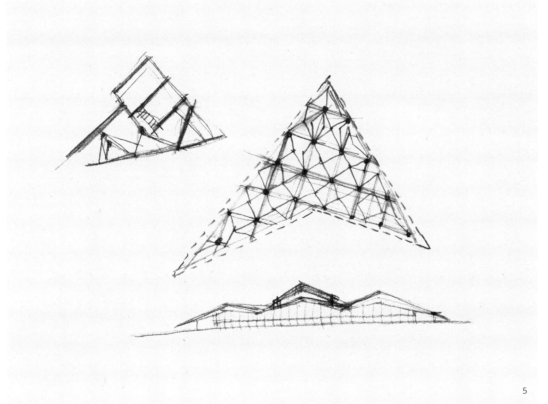

5

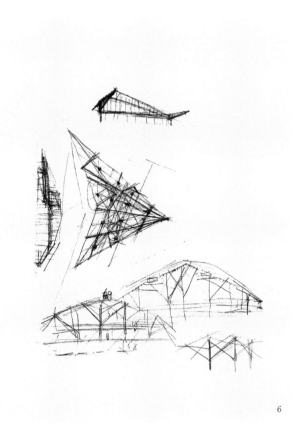

6

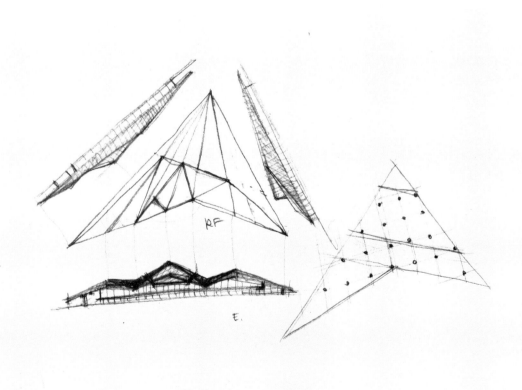

7

255

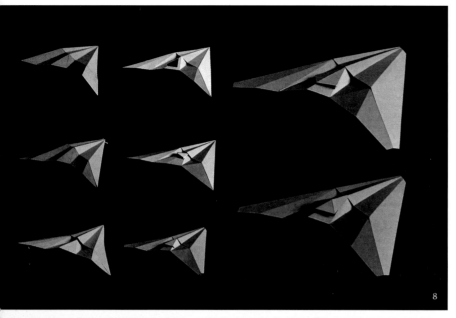

8

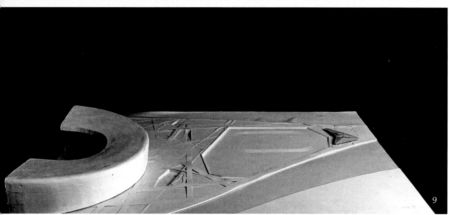

9

10

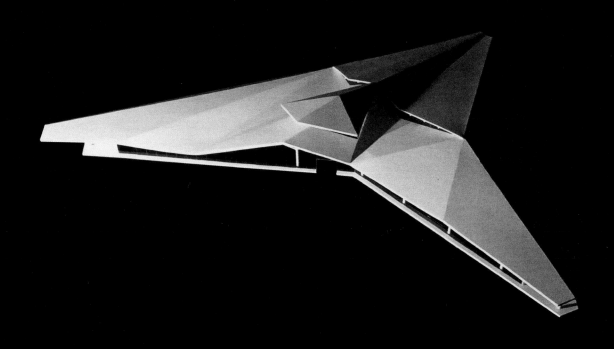

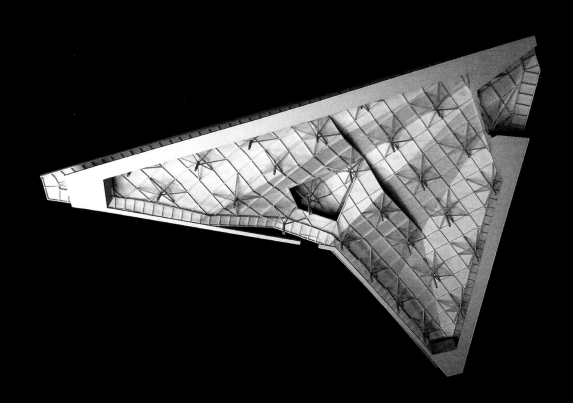

12

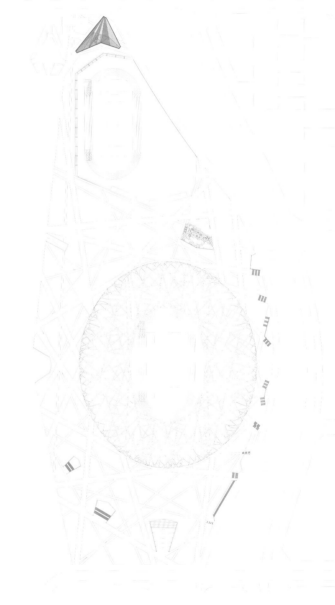

0 10 20   50m

13

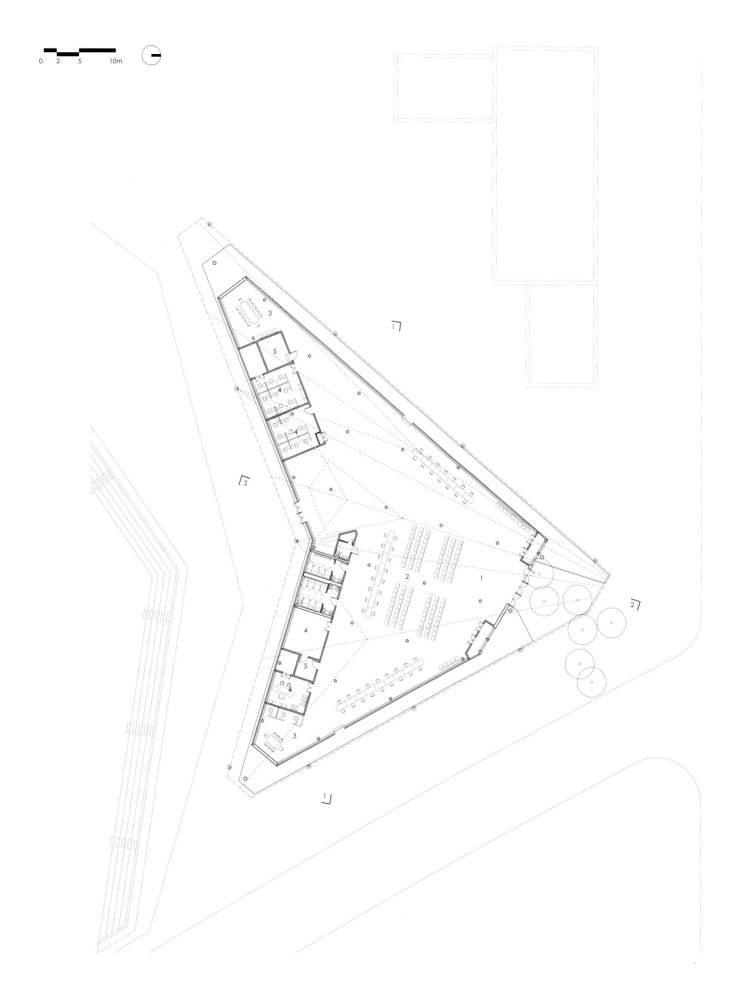

14

15

1 入口门厅 / Entrance hall
2 注册大厅 / Registration hall
3 会议区 / Meeting area
4 办公室 / Office
5 设备机房 / Equipment room

0 2 5   10m

18

21

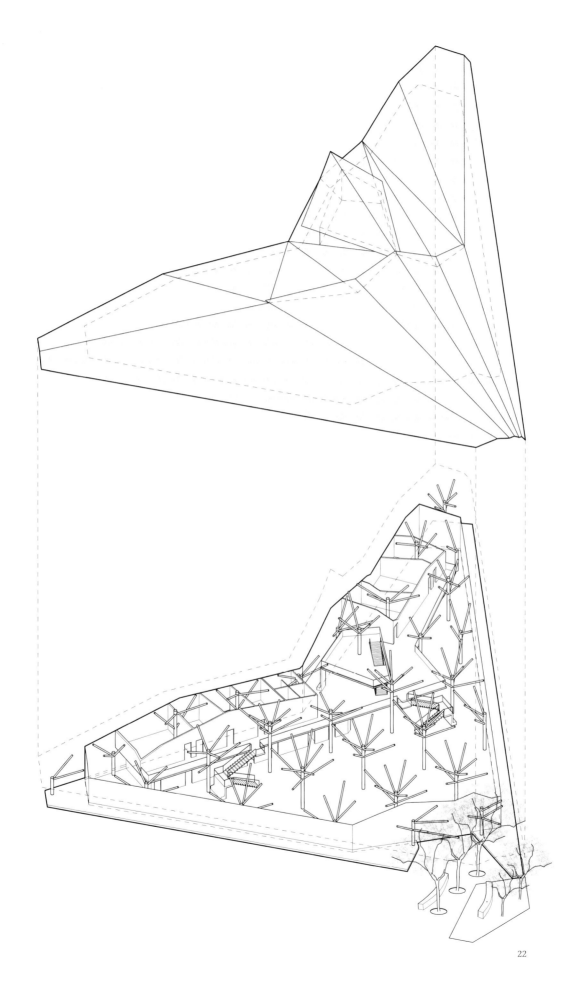

22

2015

天津大学新校区综合体育馆
GYMNASIUM OF
THE NEW CAMPUS OF
TIANJIN UNIVERSITY

天津大学新校区综合体育馆设计始于 2011 年 2 月，2015 年 11 月建成投入使用，位于天津海河中游海河教育园区里的天津大学新校区，包括各类室内运动场馆及教学、科研所需配套设施用房，以及室外田径场、各类球场、集中器械场地及极限运动场所，以满足日常体育教学、科研及师生体育锻炼需求，是一个室内与室外、地面与屋面一体的"运动综合体"。设计采用一种紧凑高效的空间组合布置方式，在紧张的用地内一体化地营造一个包含室内外多种运动功能的大学体育中心。

将各类运动场馆空间依其平面尺寸、净高及使用方式，以线性公共空间串联。一系列使用于屋顶和外墙的筒拱、直纹曲面、锥形曲面的混凝土壳体结构单元及其组合，带来适宜跨度的运动空间、高侧采光和热压通风，在外形成沉静而多变的建筑轮廓，在内露明木模混凝土筑造肌理，达到建筑结构、空间与形式完美统一的效果，犹如多簇运动空间组合而成的密集"聚落"。在几何逻辑控制下对建筑结构 / 空间单元的探寻和运用，建立起建筑及其空间自立的存在感和场所感，与由于快速建设而呈现断裂与空白的新校区场地"自然"呼应对话；同时，空间中结构的不同尺度和形状呼应着人的身体及其运动所产生的不同延伸状态，形成强烈的空间场域并唤起一种使人沉浸其中的特定情境。采用自然通风、自然采光、自然排水的系统化绿色建筑设计方式，有效节省能源，实现良好的环境和经济效益。相比于时下通过装饰获得的夸张、恣意的建筑形象，本设计中裸露的结构产生的"建构"之美，以及由此带来的沉静、朴素和富有韵律感的空间，呈现的则是另一种更加恒久的空间"诗意"。为此，建筑师在结构设计和施工、工法、精度控制等方面进行了大量主导性工作。

建筑外部材料主要采用清水混凝土饰面结合具有天津大学老校区建筑特色的深棕红色页岩砖拼贴饰面；室内各运动空间除露明本色混凝土肌理（墙柱和屋顶）及白色涂料（墙面和吊顶）的部位外，还采用了具有吸音功能的本色木丝板材墙面和结合了空调、风道、风口的本色欧松板材固定座椅，以增加空间的温暖感和舒适性。

摄影 / Photographer:
张广源、孙海霆、张虔希 / Zhang Guangyuan, Sun Haiting, Zhang Qianxi

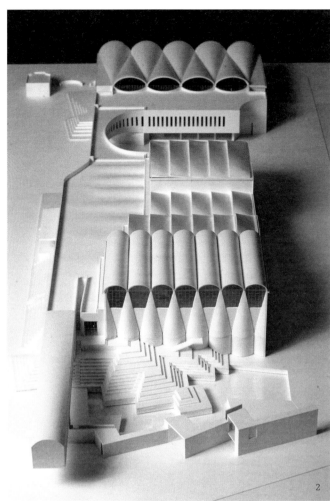

The Gymnasium of the New Campus of Tianjin University was initially designed in February 2011 and put into use in December 2015. It is located at the northern end of the new campus of Tianjin University at the Haihe Education Park in the middle stretch of the Haihe River, Tianjin City. The building's main uses include a variety of indoor sports venues, rooms for teaching and scientific research, outdoor track and field, all kinds of courts, centralized equipment fields, and extreme sports venues. As a sports complex that integrates indoor sports activities with outdoor activities, the goal of the project was to meet the daily needs of sport, including teaching, training and researching for teachers and students. The design adopts, in a compact and efficient way, a space and layout appropriate for a multi-use university athletic center within a tight land area.

A linear public space was designed to connect different sports venues in accordance with each activity's rules regarding court size, spatial boundaries, and equipment. A series of structural units, including cylindrical arches, ruled surfaces, and coned concrete arches, were integrated to support the ceiling and serve as the building's exterior walls. Using this combination of shapes and forms, we were able to create a vast indoor space for athletic activities, provide unobtrusive lighting from above, and regulate ventilation using the heat-press method. The result is a perfectly unified structure of space and function. It has varied architecture outlines externally and exposed texture of timber board-formed concrete internally. The building looks like a dense settlement composed of multiple sports spaces. This more logical and geometric application and expression of architectural structures establishes a sense of place for the building that generates a dialogue with its disjointed and indistinguishable natural surroundings caused by the rapid construction. Along with its relationship to its environment, the different scales and shapes of the structure echo the extension and movement of the human body immersing in their activity. Green building strategies such as natural ventilation, allowing for sunlight, and natural drainage are used systematically to save energy effectively while generating noteworthy economic and social benefits. In contrast to the current trends of ostentatious and arbitrary architectural adornments at one extreme and the fashionable "light" architecture at the other extreme, the gymnasium highlights the beauty of tectonics and strength of exposed concrete. It presents a calm, subtle, rhythmic, and poetic space. To this end, the atelier developed innovative structural design methods to ensure the materials and construction processes followed precisely controlled requirements.

The main exterior materials of the building are natural concrete and dark-red shale bricks, the latter of which is a characteristic material used in the old campus of Tianjin University. The indoor sports spaces mainly exposed concrete with natural color texture (wall-column and roof) and whitewash (wall and ceiling). The sound-absorbing unpainted wood-wool panel wall and the unpainted oriented strand board (OSB) fixed chairs combined with air conditioning, air ducts and air vents were also designed to increase the sense of warmth and comfort in the interior.

3

4 平面和立面草图 / Plan and elevation sketches
5 体块草图 / Volume sketch
6 立面草图 / Elevation sketches
7-8 细部草图 / Detail sketches

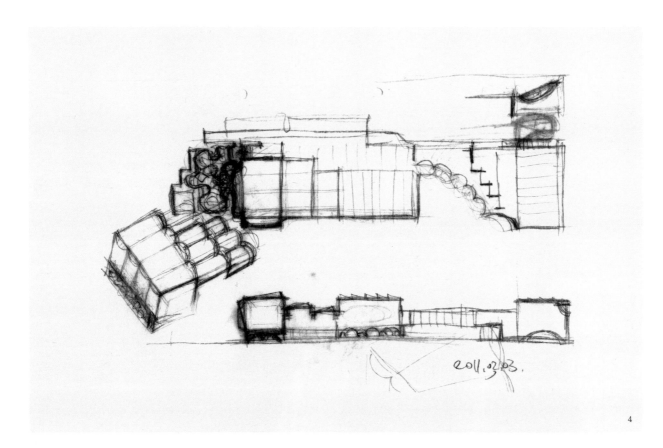

4

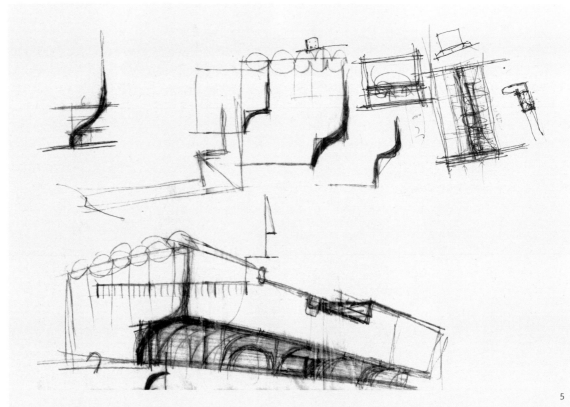

5

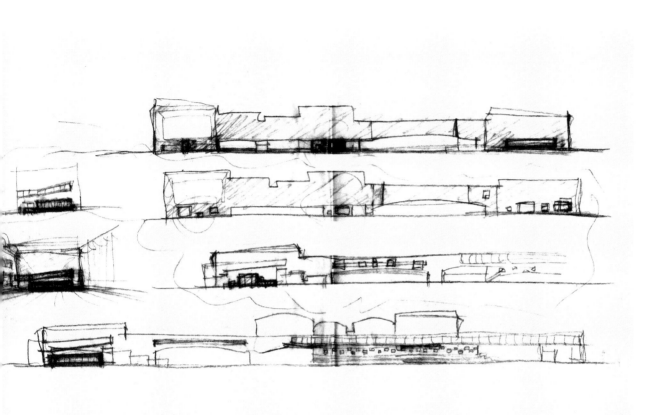

6

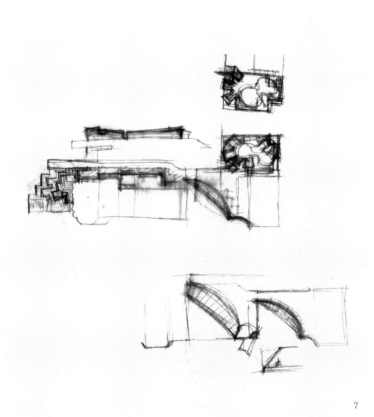

7

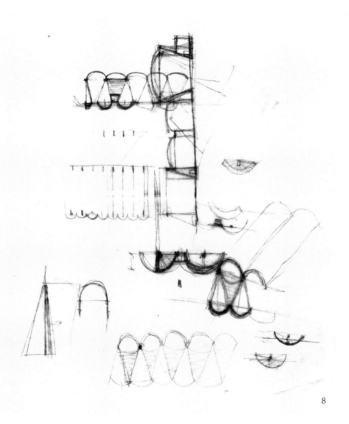

8

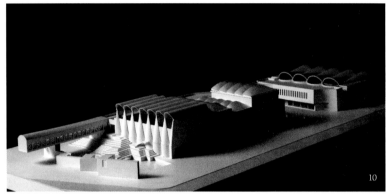

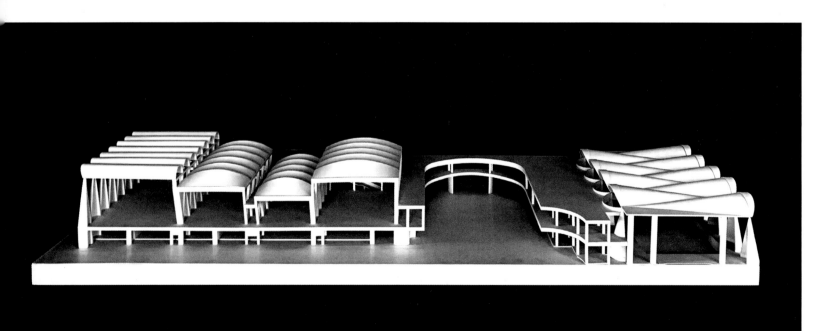

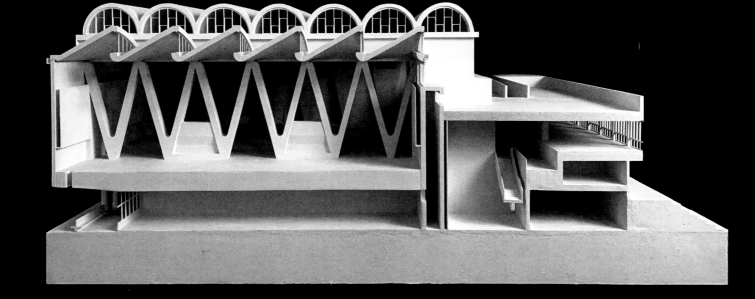

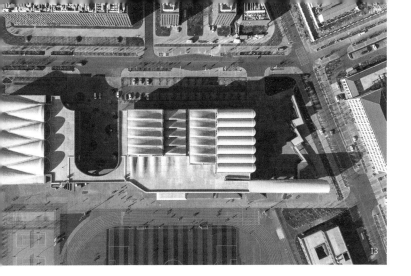

13

14

0 10 20   50m

15

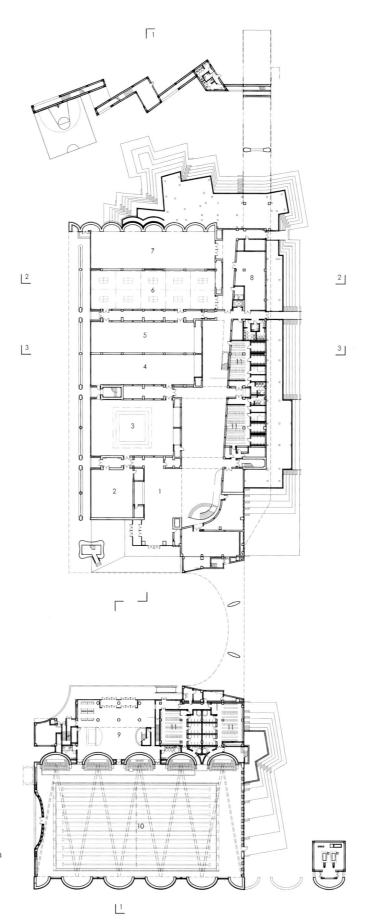

1 入口门厅 / Entrance hall
2 报告厅 / Lecture hall
3 跆拳道厅 / Taekwondo hall
4 体操室 / Gymnastics room
5 健身房 / Gym
6 乒乓球厅 / Table tennis hall
7 舞蹈室 / Dancing room
8 机房 / Equipment room
9 游泳馆门厅 / Entrance hall of the natatorium
10 游泳馆 / Swimming hall
11 更衣室 / Dressing room

16

0 5 10 20m

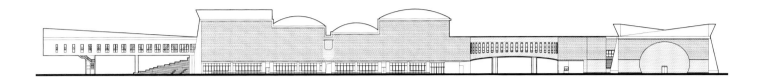

17

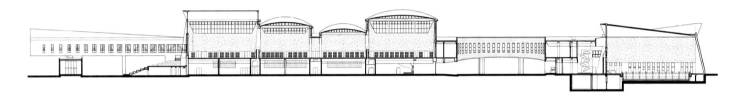

18

19

21

20 体育馆公共大厅屋顶 / View of the roof of the entrance hall
21 体育馆公共大厅上层空间 / Interior view of the entrance hall on the second floor

279

0 5 10   20m

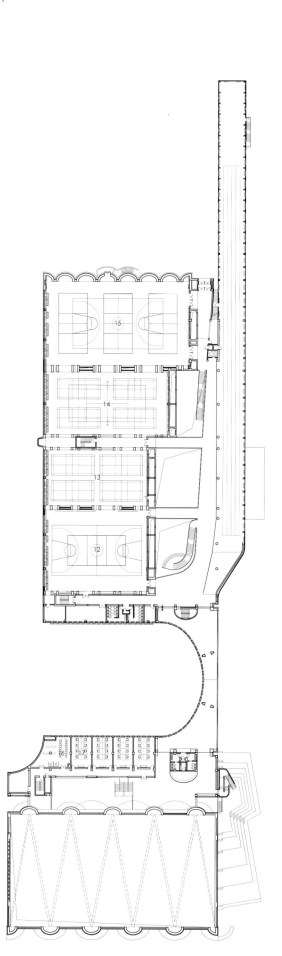

22

0 2 5 10m

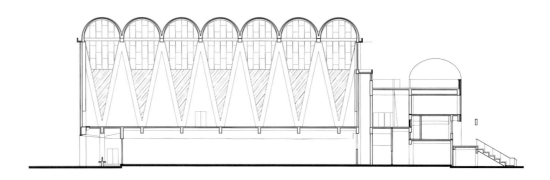

26

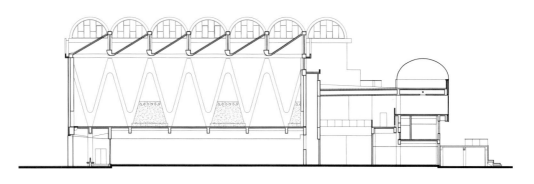

27

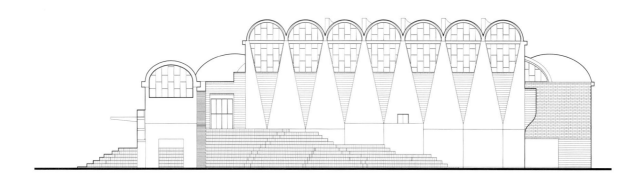

28

31 南立面人视 / View of the south facade
32 游泳馆室内 / Interior view of the natatorium

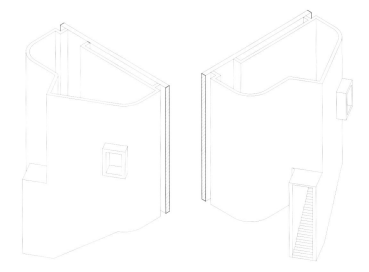

35

34

36

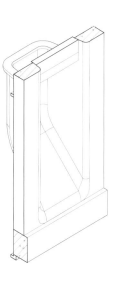

37

38 施工中的北立面 / North facade under construction
39 施工中的锥形曲面混凝土壳体结构 / The concrete structure
under construction
40 总体轴测图 / The axonometric drawing

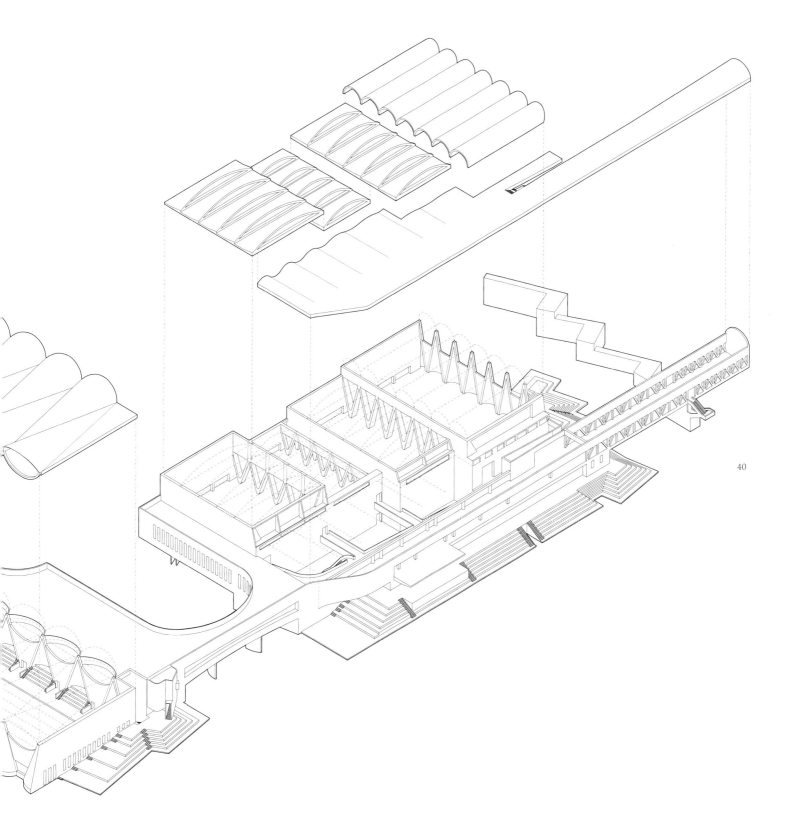

40

2015

元上都遗址博物馆
MUSEUM FOR
SITE OF XANADU

元上都遗址博物馆设计始于 2009 年 5 月，2015 年 12 月竣工，是配合元上都古城遗址申报世界文化遗产的配套项目，主要功能空间包括展厅、观众服务、藏品库房、内部办公、考古科研等。元上都遗址向南 5 公里，有一座平地隆起的草原山峰，名"乌兰台"，相传是当年忽必烈为拱卫元上都而在此设置的烽火台之一，山顶有一座巨大的敖包，由当地牧民长年累月以块石堆垒而成，蓝色的哈达随风飘扬，登上乌兰台顶，顿觉天地的宽广在眼前平铺延绵，而遗址城垣的人工矩形携着巨大的尺度，让人情不自禁地感受自然的广袤永恒和王朝的兴衰变迁。博物馆即选址于乌兰台东侧面向遗址方向的半山腰处，参观者由南而来，绕山而行，通过东北侧山脚下的道路进入博物馆区，有隐藏而豁然出现之感。

设计结合并充分利用现状废弃的采矿场来布置博物馆的建筑主体，以修整因采矿而被破坏的山体。供博物馆工作人员使用的入口设置在现状的一处折线形采矿条坑南端，并将办公、考古科研用房沿折线凹地布置，且沿山坡形状覆土；保留另一处现状圆形矿坑，经修整作为博物馆的下沉庭院，观众服务区环绕着此庭院。遵循对文化遗产环境完整性的最小干预原则，将大部分建筑体量掩藏在山体之内，仅半露一小段长条形体，隐喻遗址的城垣，将其由正北向东旋转 18 度，与山体等高线相交，并指向都城遗址中轴线上的起点——明德门，使建筑对遗址有理想的视角和轴线关联；而由明德门处看遗址博物馆，建筑则缩为一个隐约的方点，体现出对遗址环境完整性的尊重以及人工与自然的恰切对话和协调。

沿着博物馆的内外参观路径设置了一系列远眺遗址和草原丘陵地景的平台，直至到达山顶敖包，长长的路径和连缀其上的平台是博物馆不可分割的组成部分，将元上都的历史、文化和景观在此串联。

为呼应"乌兰台"在蒙语中 "红色山岩之上的烽火台"之意，建筑主体的外墙和平台、挡墙都采用了一种掺氧化铁骨料的清水混凝土，使外露的建筑体量呈现出一种斑驳的红色，犹如从山体中延伸而出，与四季变化的草原丘陵相应和，呈现出广袤与苍凉的场所气质。

摄影 / Photographers:
张广源、李兴钢 / Zhang Guangyuan, Li Xinggang

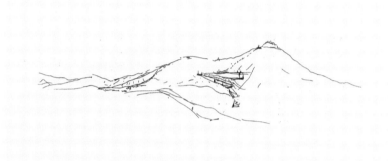

The Museum for Site of Xanadu was initially designed in May 2009 and put into use in December 2015. As a supporting project for the Site's declaration as a World Cultural Heritage in conjunction with the site of Xanadu, the ancient capital of Yuan Dynasty, its main parts include an exhibition hall, visitor service area, a collection warehouse, administrative offices, and facilities for archaeological research. Five kilometers south of the Ancient Capital of Yuan Dynasty Site, a peak rising from the flat grassland named Wulantai was said to be one of the beacon towers set up by Kublai Khan in order to protect Xanadu city. A large Oboo atop the hill was built by local herdsmen with blocks of stones for many years. The blue hada flutters in the wind. On the top of the Wulantai, one can feel the vast heaven and earth spreading horizontally in front of you. The large man-made rectangular structures and walls are the remnants of the rise and fall of a great dynasty. Visitors cannot help but feel small amidst the vast, eternal natural landscape. The museum facing the ruins is located halfway up the hill, on the east side of Wulantai. Visitors come from the south and walk around the hill. They enter the museum area through the road at the foot of the northeast side of the hill. The museum previously hidden in the hill suddenly appears, surprising visitors.

The main body of the museum was embedded into an existing abandoned mining site, repairing the hill damaged by the mining. The museum's staff entrance is set at the southern end of an existing mining trench. The office and archaeological research rooms are arranged along a crease along the shape of the hill. Another existing circular pit is preserved, and transformed into a sunken courtyard surrounded by the visitors' service area. Following the principle of minimal intervention to preserve the integrity of the environment of cultural heritage, a majority of the building is hidden into the hill with only a small long strip exposed. It functions as a metaphor for wall ruins. The architects rotated the strip from north to east by 18 degrees to intersect with the hill contour. Pointing to the starting point—Mingde Gate on the central axis of the capital city site, the building has an ideal perspective towards and axial relation with the site. Viewed from Mingde Gate, the building is reduced to a subtle point, reflecting the respect for the environmental integrity of the site and the proper intention of dialogue and coordination between man and nature.

Along the museum's interior and exterior visiting paths linking to the peak, a series of platforms upon which people may overlook the ruins and hilly grassland are set up. The long paths and platforms are an integral part of the museum, connecting history, culture, and the landscape.

To echo the meaning of Wulantai in the Mongolian language—the beacon tower above the red rock, the exterior walls, platforms and retaining walls of the main building are made of bare concrete mixed with iron oxide aggregates. The exposed building exterior presents a mottled red color, as if emerging from the mountain and blending in to the hilly grassland as the seasons change. It presents a vast and desolate temperament.

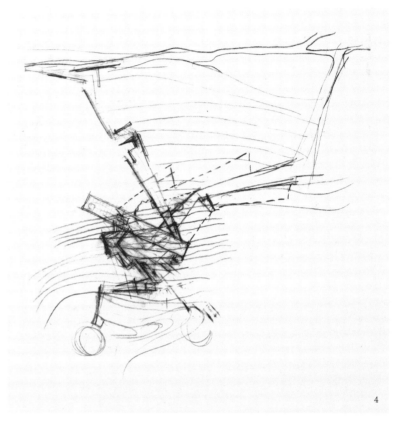

4

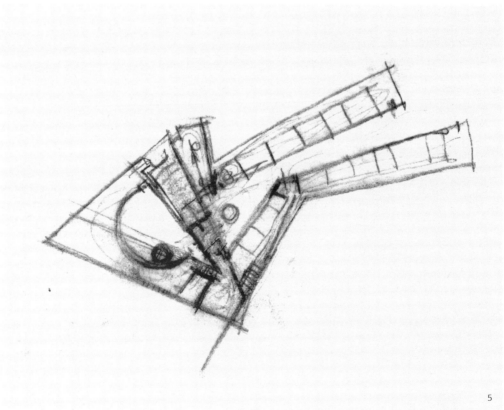

5

6

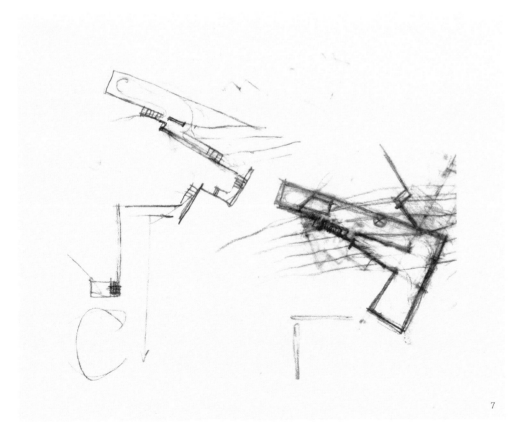

7

8

0 10 20   50m

9

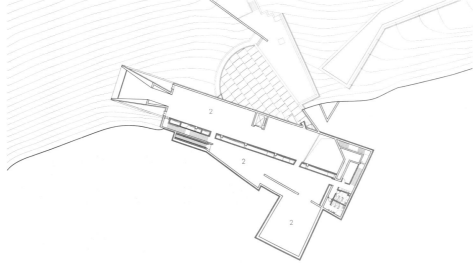

11

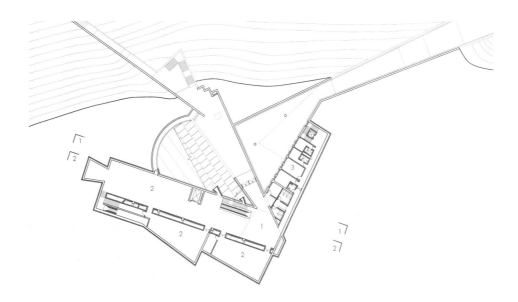

12

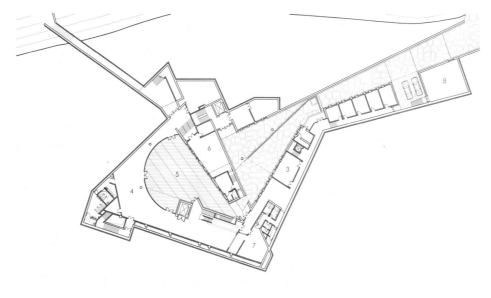

1 入口门厅 / Entrance hall
2 展厅 / Exhibition hall
3 研究室 / Research room
4 餐厅 / Restaurant
5 庭院 / Courtyard
6 报告厅 / Lecture hall
7 厨房 / Kitchen
8 机房 / Equipment room

13

14 剖面 1-1/ Section 1-1
15 剖面 2-2/ Section 2-2
16 东立面 / East elevation
17 东侧人视 / View from the east

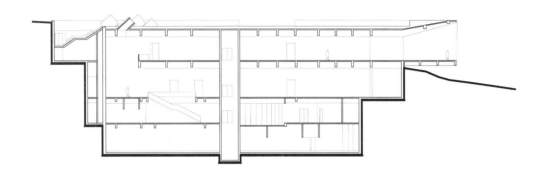

14

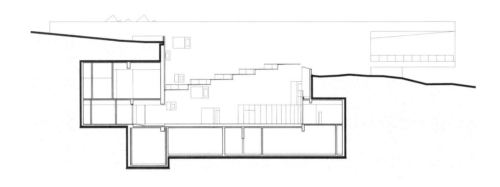

15

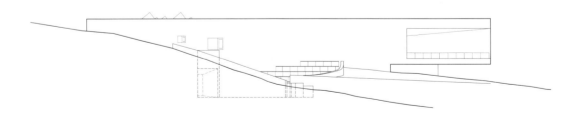

16

18

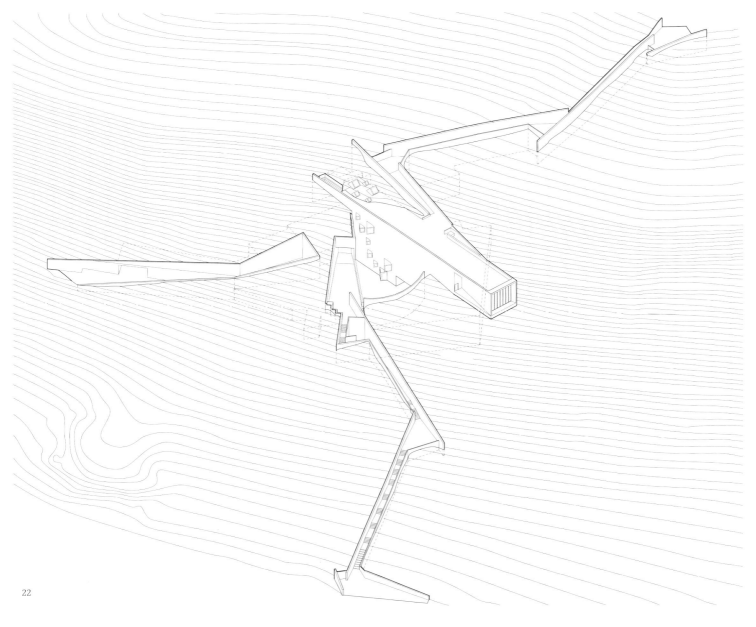

22

2016

北京朝阳区生活垃圾综合处理厂焚烧中心
WASTE TREATMENT CENTER IN CHAOYANG

北京朝阳区生活垃圾综合处理厂焚烧中心设计始于 2011 年 2 月，2016 年 10 月建成投入使用，位于北京市朝阳区金盏乡高安屯村生活垃圾综合处理厂区，是一座拥有国际先进设备及工艺的现代大型垃圾焚烧中心。垃圾是城市生活的必然产物，随着城市的快速发展，垃圾处理中心也越来越成为城市重要的市政公用设施，需要改变其刻板单一、与"污染"关联的"厂房"形象，传递健康可持续生活方式的多样需求，以更亲和的姿态承担宣传教育的社会公共责任。

焚烧中心主厂房是厂区内体量最大、高度最高的建筑单体和核心建筑，生活垃圾在此完成"废弃物——焚烧减量——无害化处理——产生能源"的循环再生过程。通过对内部工艺和外部围护系统的整合设计，建筑体量满足设备工艺对空间的要求，使建筑作为一个整体，将垃圾运卸、存放、焚烧、净化、发电的工艺流程及空间特征表达出来，并在保证工艺要求和使用功能的前提下，强调建筑体量的完整性和工业建筑的力量感、尺度感。主厂房内部各个空间对物理环境的不同需求也直接反映在建筑外立面上，选用能够反映工业建筑特征的立面建材，镀铝锌圆浪型截面波纹钢板竖向纹理排布，充分利用材料搭接安装的构造特点有效隐藏拼缝，增强整体感，形成简洁、朴素、高效、不过分夸张喧闹的外观。在方整的基本形态基础上，在建筑转角、主要出入口、屋顶等位置进行了体量的细化处理。在建筑底部 6 米标高范围内形成了一个连续的"基座"，整合上部尺度不一的矩形体量，构成一个内部联通的空间整体。主厂房南端 80 米高的烟囱是整个园区的制高点，采用区别于常规烟囱的四心椭圆断面，并在顶部做切削处理，其抽象挺拔的造型也与方整的主体建筑形成对比。在具备工业建筑的尺度感、力量感、雕塑感的同时，与常规的厂房拉开距离，也不同于一般的"公共建筑"，从而获得自身独有的气质。

由于焚烧中心还具有展示、教育的社会服务功能，建筑师在厂房内部空间设置了一条生动、安全、清洁、全面的教育展示通廊和一条垃圾处理参观廊道，参观和展示区域与内部生产区域流线分开，互不干扰，并重点处理了办公、参观、展示和教育空间的立面表达。

摄影 / Photographer:
张广源 / Zhang Guangyuan

The Waste Treatment Center in Chaoyang, Beijing, was initially designed in April 2011, and completed and put into use in October 2016. It is a modern, large-scale waste incineration center, using internationally advanced equipment and technology. Garbage is an inevitable byproduct of urban life. With rapid urban development, garbage treatment centers have become an increasingly important municipal facility for the city. What is needed is to change its stereotyped image of a factory building associated with pollution. These treatment centers are part of the diverse needs of a municipal government to maintain its citizens' healthy and sustainable lifestyle. With a more approachable design, waste treatment centers can assist in promoting education, social responsibility, and public health.

The main plant of the incineration center is the core building with the largest volume and height in the plant area. The sustainable aspects of waste processing–incineration reduction, treatment, and energy generation–is completed at the main plant. The building's internal space accommodates all the processing equipment by integrating the internal processes with the exterior, enclosed system. The building expresses the process flow and spatial characteristics of garbage transportation, storage, incineration, purification, and power generation as a whole. While ensuring that the function of the center meets processing requirements, the design highlights the strength, integrity, and scale of the treatment center. The building facade clearly reflects the function, uses, and industrial character of the spaces inside. The architects selected a cladding of corrugated steel plate with aluminum-plated zinc circular waves as the primary exterior material. The plates are arranged vertically to conceal seams to present an organized, quiet appearance. Such an approach enhance the building's comprehensive goals reflecting its character as an industrial building, and presenting itself as a calm, efficient, and modest structure. Based on its basic square form, the building volume is refined through detailed design of the corners, main entrance and exit, and the roof. The building sits atop a continuous 6-meter high pedestal, which integrates various rectangular structures of different sizes into an internally coherent whole. The 80-meter high chimney at the south end of the main power house is the commanding point of the whole park. It adopts a four center elliptical section different from conventional chimneys and is clipped at the top. Its abstract and straight shape contrasts with the square main building. Because of its scale, power, and industrial sculptural characteristics, the Waste Treatment Center forges a unique identify amidst other public buildings and factory buildings.

The incineration center also has a social service function for visibility and education. A dynamic, safe, clean, and comprehensive educational exhibition corridor and a garbage disposal visiting corridor are set up in the interior space of the factory. The specially designed visitor and exhibit areas are separated from the internal production area so as not to disturb either's function.

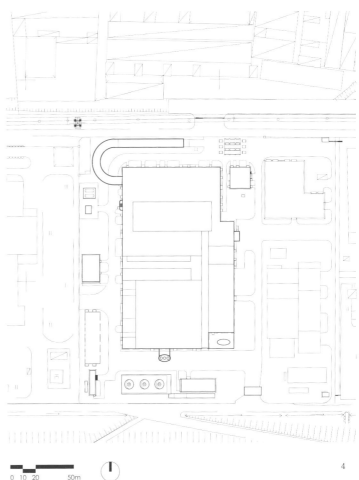

4

0 10 20　　50m

3

5 货运架空廊道 / View of the overhead channel
6 西南侧人视 / View from the southwest

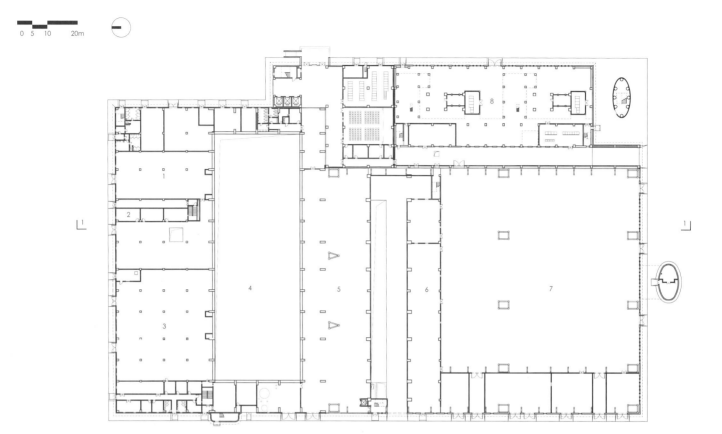

1 空压机站 / Air compressor station
2 办公室 / Office
3 化水车间 / Melting workshop
4 卸料车间 / Unloading space
5 焚烧间 / Burning space
6 配电室 / Distribution room
7 烟气净化间 / Purification space
8 汽机间 / Turbine room

7

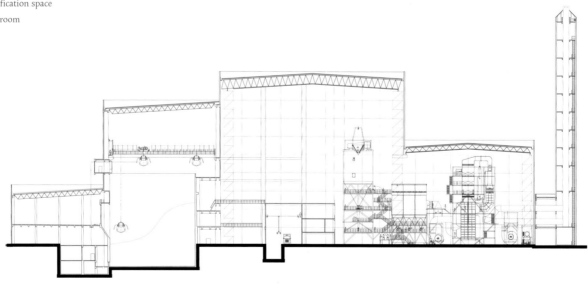

8

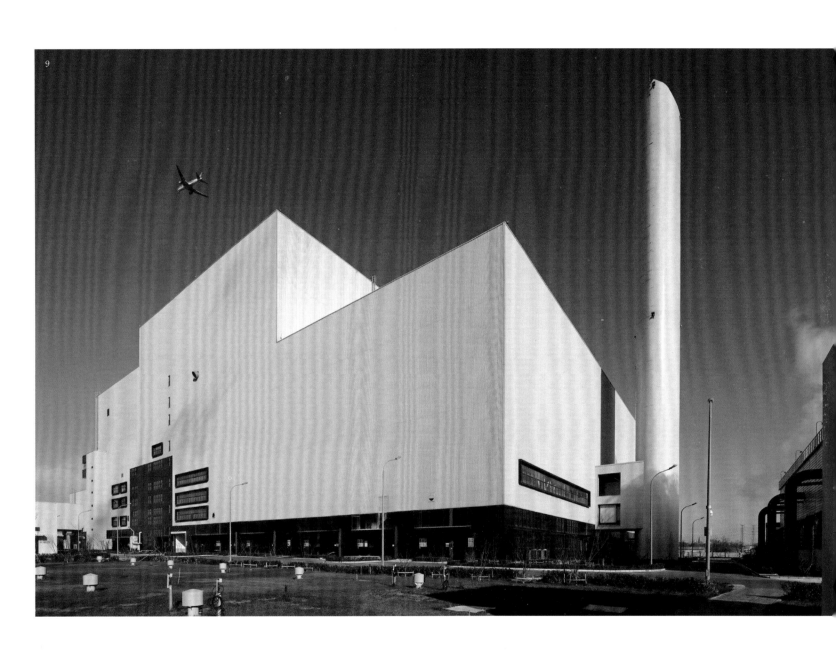

2017

通辽美术馆和蒙古族服饰博物馆
TONGLIAO ART GALLERY & MONGOLIAN COSTUME MUSEUM

通辽美术馆和蒙古族服饰博物馆设计始于 2014 年 3 月，2017 年 9 月竣工，是具有科尔沁文化特征的功能复合型文化公园中的两个公共项目，位于通辽市孝庄河岸边的狭长绿地内，美术馆在东侧北岸河道转弯狭窄处，服饰博物馆位于西侧北岸最宽处。时间发展、历史更迭、文化进步，而自然和气候特征始终维持在一个稳定的状态，使得不同的地域及人群的特征具有了识别性。蒙古族人的气质深沉厚重，而其人造物（如蒙古包、勒勒车）则便携轻盈，服饰博物馆和美术馆分别对应"厚重"和"轻盈"，呈现出蒙古族文化的不同侧面，并采用了相似的几何构形单元和空间体验模式。

美术馆西北侧紧邻城市道路，南侧紧邻河岸，用地狭长弯曲，首层建筑体量采用覆土的方式隐藏在一个小丘之下，并局部切削下沉形成入口广场；双筒状长线型的主体建筑体量轻轻置放在山丘之上，两端微微上翘，形成单纯而富有弹性的形体曲线，悬挑凌驾于河湾之上，指向河道的东北和西南两端，中间轻触大地，呈现出一种最小限度介入自然的轻盈姿态；采用了钢筋混凝土剪力墙和钢结构组合体系构成平行并渐变的"复廊"空间，两端部形成不对称的双拱组合立面及室外景观平台，中间双墙作为主要悬挑结构，同时也是垂直交通和设备空间；建筑采用覆土种植屋面、清水混凝土、直立锁边金属幕墙等材料及做法以凸显建筑的"轻盈"状态。

服饰博物馆处在河道北岸由阔至窄的区域，采用三个相同尺寸的体量呈放射状分别面向城市、河流和入口广场；建筑主体下沉半层，可由三个方向进入建筑内部，面向河岸的单元作为观演厅和公共活动空间，展示空间地面标高阶梯状升起，串联形成连续环状参观流线；三组高低组合连续渐变的混凝土马鞍拱形薄壳单元整体浇筑在一起，曲线截面舒展开阔，形成空间的张力，隐喻了蒙古族服饰的厚重特征；每个方向的薄壳单元在端部开放并形成双拱组合立面及室外平台和公共空间，呼应和框界不同方向的内外景观，三个单元交汇的地方作为公共交通空间，中央上方采用半透明的膜结构把三个薄壳屋顶连接起来。主要采用碎拼彩釉面砖屋面和墙面，局部采用钛锌板直立锁边金属屋面。

摄影 / Photographer:
李兴钢、梁旭 / Li Xinggang, Liang Xu

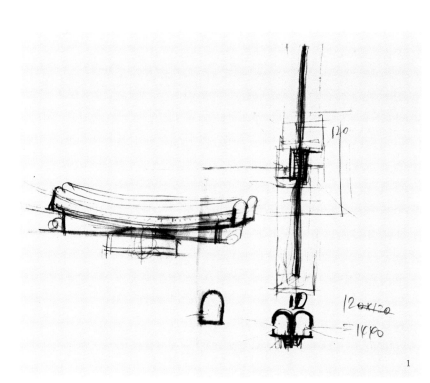

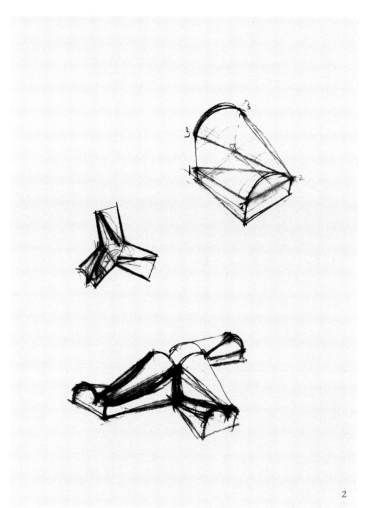

1

2

The Tongliao Art Gallery & Mongolian Costume Museum are initially designed in November 2013, and completed in September 2017. The two public projects were built in a cultural complex to feature the cultural heritage of Horqin culture. The gallery and museum are located on a long, narrow lawn on the banks of the Xiaozhuang River in Tongliao City. The art gallery is located on a narrow bend of the river on the eastern side, whereas the costume museum is located in the widest portion of the western side of the river's north bank. In spite of the progress of time, changes in history, and the progress of culture, nature and climate remain relatively stable, which makes the characteristics of different regions and human beings distinctive. The temperament of Mongolians is strong and stable, while their creations (such as yurts and rickshaws) are portable and light. The design of the costume museum and art gallery, with similar geometric elements and spatial experience patterns, corresponds to heaviness and lightness respectively, showing different aspects of Mongolian culture.

The northwest side of the art gallery is close to the city road, while the south side is close to the river bank. The site is narrow, long and curved. The building's first floor is hidden under a hill and covered by earth. It is partially cut and sunken to form the entrance square. The dual cylinder long linear space is gently placed on the hill. The two ends are slightly upturned, forming a simple and elastic, curved shape overhanging over the river bay. On the top, it points to the northeast and southwest ends of the river. The building touches the earth in the middle, showing a light posture with minimal intrusion into nature. The composite structural system of a reinforced, shear concrete wall and steel arches are used to form a parallel and gradually changing corridor space. An asymmetric double arch composite facade and outdoor platform appear at the two ends. The middle double wall serves as the main cantilever structure and provides vertically movement and equipment space. The building adopts the methods of turf roofs, pale concrete and an upright metal curtain to highlight a sense of lightness.

The costume museum is located in the area that gradually narrows on the north bank of the river. Three volumes of the same size are oriented radially relative to the city, river and entrance square. Visitors can enter the building from three directions. The main body of the building is recessed a half floor. The unit facing the river bank serves as the performance hall and public space. The height of the exhibition space rises gradually, creating a continuous circular visiting circulation. Three sets of thin, concrete saddle arched shell units with alternating height levels are poured together as a whole, creating an extended and wide curved section. This spatial tension becomes a metaphor for the heavy characteristics of Mongolian clothing. Each thin shell unit is characterized by the double arched facade, which has an open platform and public space. These shell units connect and frame the internal and external landscape in different directions. The area where the three units meet becomes a public transportation space, covered by a translucent membrane structure above the center. Broken, colored glazed tiles adorn the roof and walls. Portions of the roof are formed by titanium zinc plates.

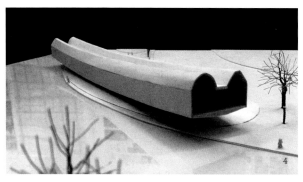

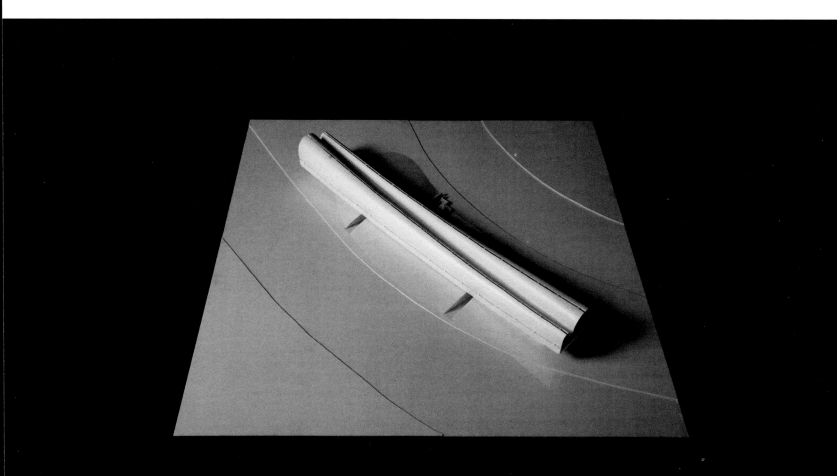

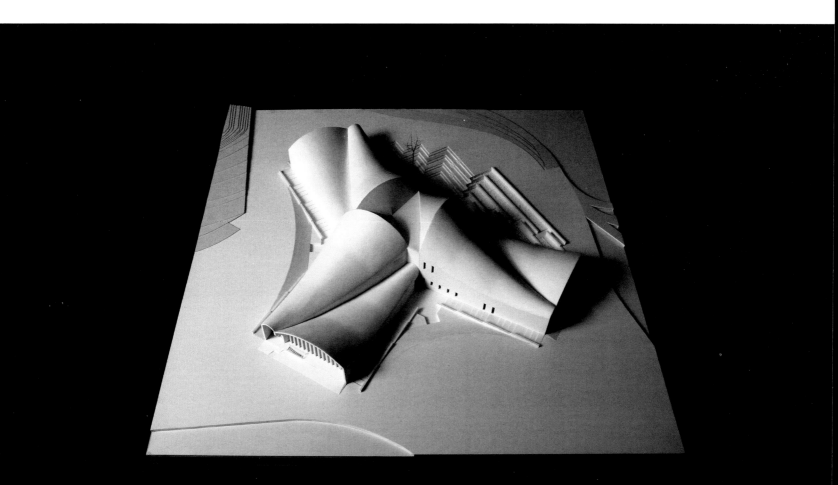

10 西北侧鸟瞰 / Aerial view from the northwest
11 总平面图 / Site plan

10

11

0 50 100   200m

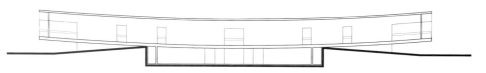

12 美术馆东北侧人视 / View of the art gallery from the northeast
13 剖面 1-1/ Section 1-1
14 美术馆一层平面 / The 1st floor plan of the art gallery
15 美术馆二层平面 / The 2nd floor plan of the art gallery

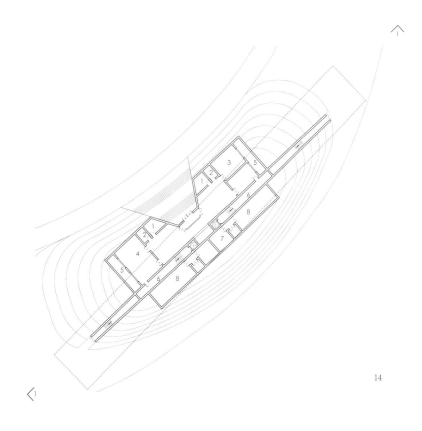

14

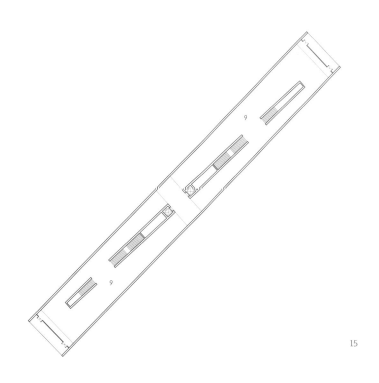

15

1 值班室 / Duty room
2 卫生间 / Lavatory
3 办公室 / Office
4 多功能厅 / Multi-function hall
5 窗井 / Patio
6 储藏室 / Storage
7 机房 / Equipments
8 库房 / Storeroom
9 展厅 / Exhibition hall

18 服饰博物馆一层平面 / The 1st floor plan of the museum
19 剖面 2-2/ Section2-2
20 服饰博物馆西南侧人视 / View of the museum from the southwest of the museum

0 2 5 10m

18

1 入口 / Entrance
2 联系厅 / Contact hall
3 售票 / Ticket
4 值班室 / Duty room
5 中厅 / Central hall
6 服务台 / Service counter
7 门厅 / Lobby
8 消防控制室 / Fire control room
9 入口厅 / Entrance hall
10 第一展厅 / The 1st exhibition hall
11 办公室 / Office
12 研究室 / Research
13 库房 / Storeroom
14 设备用房 / Equipments
15 第四展厅 / The 4th exhibition hall
16 临时展厅 / Temporary exhibition hall
17 出口厅 /Export hall
18 过厅 / Hall
19 休息敞厅 / Lounge

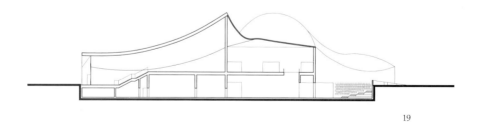

19

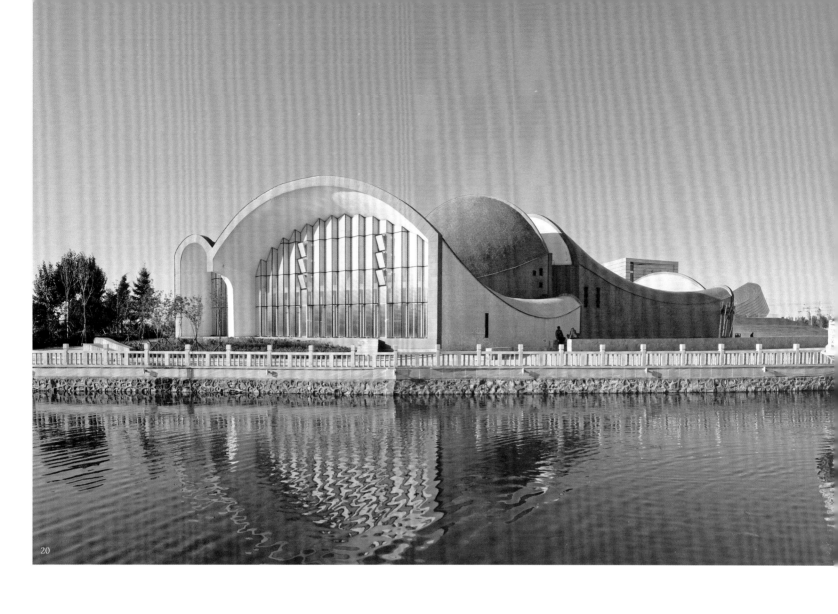

20

21

22

2017

"微缩北京"
——大院胡同 28 号院改造
"MINIATURE BEIJING",
RENOVATION OF NO.28
DAYUAN HUTONG

大院胡同 28 号院改造项目设计始于 2016 年 11 月，2017 年 9 月竣工，位于北京旧城西单—丰盛区域，将原来占地面积 262 平方米的普通杂院，在保持基本建筑外观、檐 / 脊高度不变的条件下，改造为五套带院落的居住公寓 + 一个咖啡 / 餐茶公共空间。这个微小项目是一次结合了旧城更新、院落改造、理想居所研究的设计实践：以研究传统北京的复合性城市结构为基础，认识并运用其结构可延伸、加密的特征，通过分形加密，将大杂院转变为"小合院群"；"宅园"与"公共单元"的设置适应了现代社会结构，将院落转变成"微缩社区"；居所层面极限尺度的技术性设计服务于"宅园合一"的精神性营造，通过空间叙事，将日常诗意与都市胜景的体验带入"理想居所"；以个案回应了北京旧城更新中人口密度、生活质量和风貌传承等典型难题，并探讨向更广泛的社区、城市扩展，恢复北京旧城自生机会的可能性。

由喧闹的城市商业街区转折进入闲适宁静的胡同区域，再由外部胡同通过一条半室外主巷道和一条再次分支的巷廊，分别进入北侧、南侧不同格局和规模的五套"合院"公寓——北侧三套小"宅园"，南侧一中、一大两套较大"宅园"。每套公寓拥有不同大小、形状的院庭，主要起居空间通透，面向并对景于庭园。庭园内高树之下，混凝土体块取意抽象山石，又是室外坐具，置于波形立瓦铺地之上，铺地微凹可存薄水而成水中树石之景，营造日常诗意；由主巷道继续南行，经过公共的咖啡 / 餐茶空间单元，抵达后面的公共小庭，并可沿一侧的混凝土阶梯，上至抬升在庭院上方的亭楼平台，这里是公共巷道在剖面上的延伸，视野变得开阔，夕阳西下之时，游者在此可观想、沉浸于由旧城院落的层叠屋顶、古树、飞翔的鸽群、远方城市高楼群所构成的深远胜景。

几个线型混凝土结构 / 空间单元构成了内含于整体院落群组建筑的空间架构，既内含服务空间，形成主体支撑结构，又成为联系居室和庭园的入口和廊道，并框定园景。采用了传统青瓦坡屋面、青砖外墙、花砌镂空园墙、小木模板清水混凝土墙 / 板 / 阶梯、"竹钢"木作屋顶结构 / 门窗 / 家具 / 楼梯等材料及做法。

摄影 / Photographer:
苏圣亮、李兴钢 / Su Shengliang, Li Xinggang

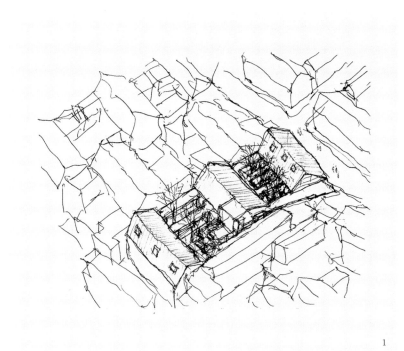

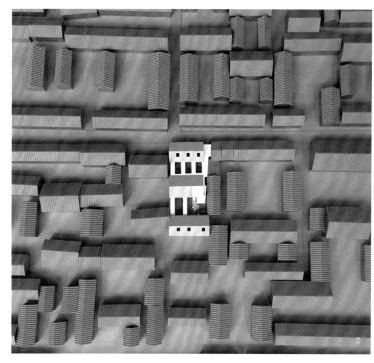

The Renovation of No. 28 Dayuan Hutong, titled "Miniature Beijing", located in the old city area of Beijing, is initially designed in November 2016, and completed in September 2017. While maintaining the basic building appearance and height of the roof line, a common courtyard house of 262-square meter was transformed into five apartments with self-contained courtyards, as well as a public space that hosted a cafe and a tea house. This small experimental project was a design practice that combined historical urban renewal, courtyard renovation and the unique opportunity to research Beijing's development as a "compound" city. Based on the study of Beijing's traditional, yet complex urban structure, along with an understanding and utilizing of its extended and densified structure, the crowded courtyard house was transformed into a collection of small courtyard dwellings. The layout of the house-garden units and a public unit fit in the structure of contemporary society, turning the courtyard house into a miniature community. The technical design of the smallest dwelling unit serves as the spiritual starting point for "house-garden harmonization". Through spatial guidance, the poetry of daily life and the urban environment are integrated into the ideal living environment. The individual case addresses three main issues when rejuvenating Beijing's ancient urban plan: population density, quality of urban life, and historical preservation. This project serves as a model for ideas that can extend to other communities and urban spaces as ancient Beijing organically evolves.

Turning from the noisy urban commercial streets into the peaceful and leisurely Hutong district, then through the outside main Hutong to a semi-external alleyway and a further smaller lane, one may see five courtyard apartments with different sizes and configurations—three small "house gardens" in the north, one middle and one large "house gardens" in the south. Each contains a courtyard that is varied in size and shape. The main living room of each apartment is spacious and bright, with the additional benefit of a garden view. Under the tall trees in the garden, the concrete block is not only an abstract mountain stone, but also an outdoor seat, placed on a tiled patio. The slightly concave of the ground can hold a thin layer of water, which reflects poetic images of the surrounding trees and stonework. Walking further south along the main alleyway, passing the cafe and tea house, visitors arrive at a small public garden at the rear. Visitors can then climb the elevated platform of the pavilion above the courtyard via a set of concrete steps. This platform is the extension of the public roadway, and the view becomes wider. With the setting sun as the backdrop, the profound scenery composed of cascading roofs of the courtyard houses in the old city, ancient trees, flying pigeons and the high-rise buildings in the distance brings visitors into a contemplative state.

Several linear concrete modules form the spatial elements within the entire courtyard complex. They not only contain service spaces and provide the primary supporting structures, but also become the thresholds and corridors that frame the garden views and connect the living rooms and the gardens. The project uses materials and practices of traditional grey tile slope roofs, blue brick exterior walls, garden walls with holes in the masonry, pale concrete wall/board/ladder with the texture of narrow timber formwork, and wooden bamboo roof structure/doors and windows/furniture/stair.

5

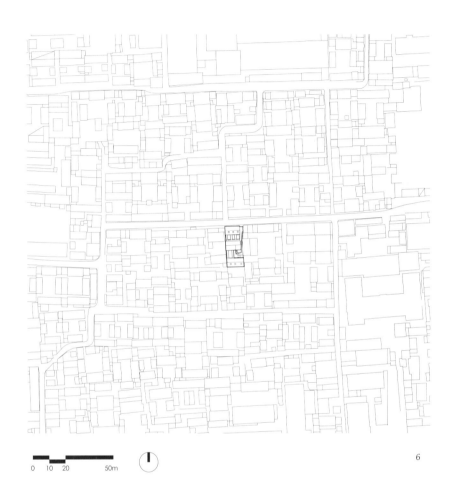

6

7

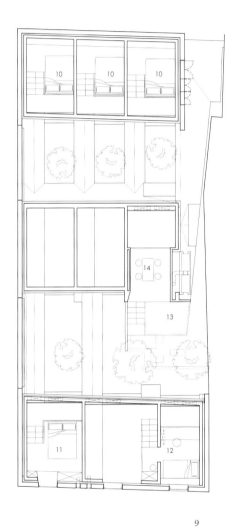

9

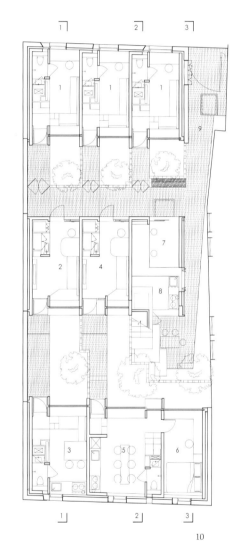

10

1 起居空间（a 户型）/ Living room of Unit a
2 起居空间（b 户型）/ Living room of Unit b
3 餐厨空间（b 户型）/ Dining room of Unit b
4 起居空间（c 户型）/ Living room of Unit c
5 餐厨空间（c 户型）/ Dining room of Unit c
6 主卧（c 户型）/ Main bedroom of Unit c
7 公共餐茶空间 / Shared dining hall
8 公共厨房 / Shared kitchen
9 主巷道 / Main lane
10 卧室（a 户型）/ Bedroom of Unit a
11 卧室（b 户型）/ Bedroom of Unit b
12 次卧（c 户型）/ Bedroom of Unit c
13 台 / Platform
14 亭 / Pavilion

11 起居空间（a 户型）/ Interior view of the living room (Unit a)
12 剖面 1-1/ Section 1-1
13 剖面 2-2/ Section 2-2
14 剖面 3-3/ Section 3-3

0   1   2   5m

12

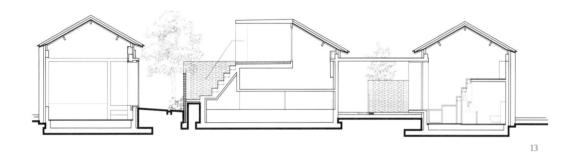

13

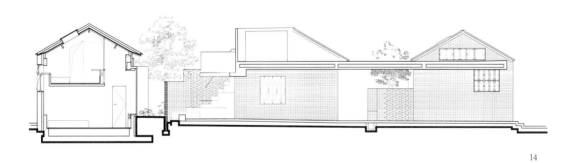

14

15 从餐厅望向内院（c 户型）/ View from the dining room to the courtyard (Unit c)
16 内院（c 户型）/ View of the courtyard (Unit c)

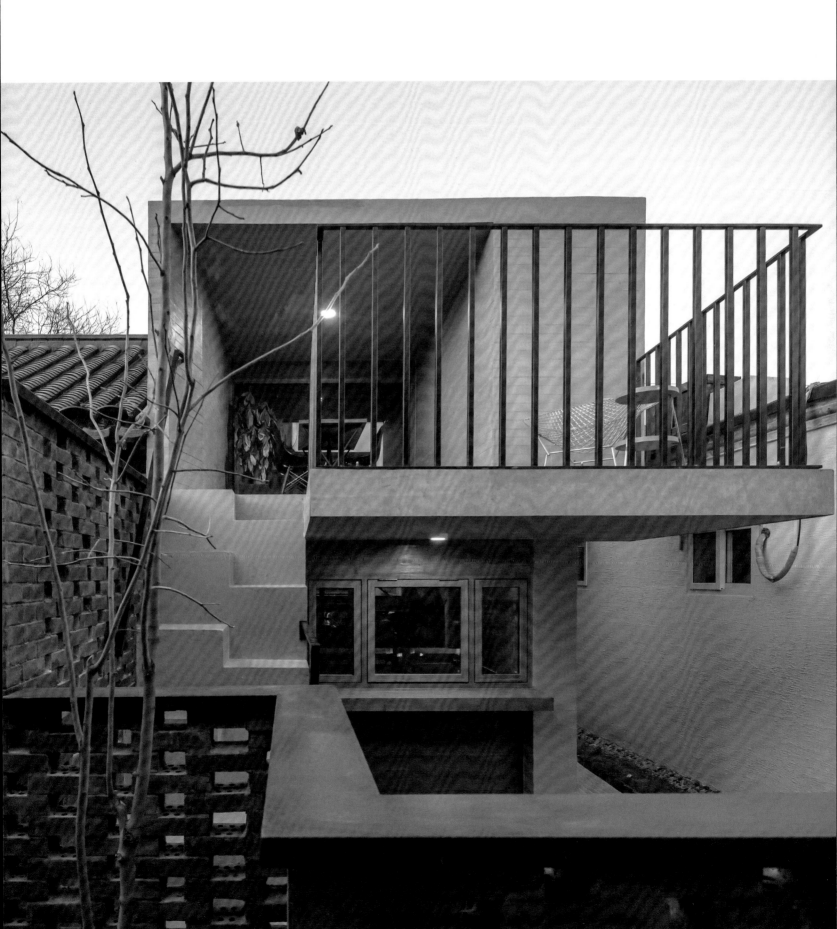

22

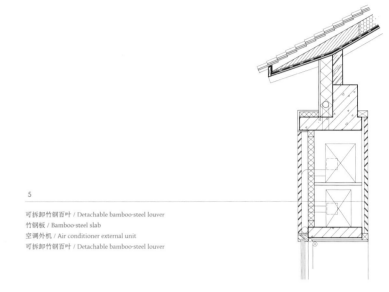

5

可拆卸竹钢百叶 / Detachable bamboo-steel louver
竹钢板 / Bamboo-steel slab
空调外机 / Air conditioner external unit
可拆卸竹钢百叶 / Detachable bamboo-steel louver

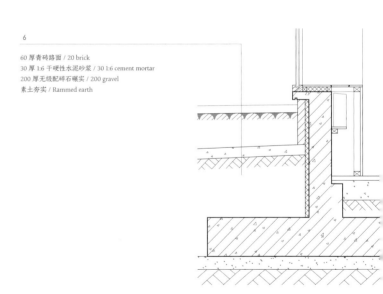

6

60 厚青砖路面 / 20 brick
30 厚 1:6 干硬性水泥砂浆 / 30 1:6 cement mortar
200 厚无级配碎石碾实 / 200 gravel
素土夯实 / Rammed earth

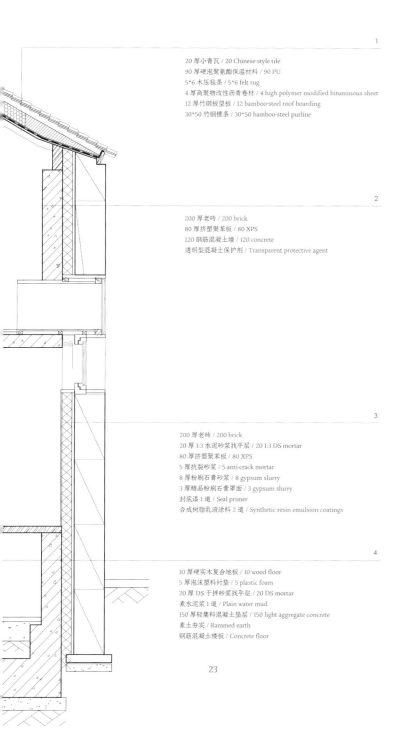

1

20 厚小青瓦 / 20 Chinese-style tile
90 厚硬泡聚氨酯保温材料 / 90 PU
5*6 木压毡条 / 5*6 felt rug
4 厚高聚物改性沥青卷材 / 4 high polymer modified bituminous sheet
12 厚竹钢板望板 / 12 bamboo-steel roof boarding
30*50 竹钢檩条 / 30*50 bamboo-steel purline

2

200 厚老砖 / 200 brick
80 厚挤塑聚苯板 / 80 XPS
120 钢筋混凝土墙 / 120 concrete
透明型混凝土保护剂 / Transparent protective agent

3

200 厚老砖 / 200 brick
20 厚 1:3 水泥砂浆找平层 / 20 1:3 DS mortar
80 厚挤塑聚苯板 / 80 XPS
5 厚抗裂砂浆 / 5 anti-crack mortar
8 厚粉刷石膏砂浆 / 8 gypsum slurry
3 厚精品粉刷石膏罩面 / 3 gypsum slurry
封底漆 1 道 / Seal primer
合成树脂乳液涂料 2 道 / Synthetic resin emulsion coatings

4

10 厚硬实木复合地板 / 10 wood floor
5 厚泡沫塑料衬垫 / 5 plastic foam
20 厚 DS 干拌砂浆找平层 / 20 DS mortar
素水泥浆 1 道 / Plain water mud
150 厚轻集料混凝土垫层 / 150 light aggregate concrete
素土夯实 / Rammed earth
钢筋混凝土楼板 / Concrete floor

23

24

357

2017

楼纳露营基地服务中心
CAMPING SERVICE CENTER IN LOUNA

　　楼纳露营基地服务中心设计始于2016年11月，2017年10月基本竣工投入使用，位于贵州省兴义市东部山区的楼纳村大冲组"建筑师公社"——群山环绕下的一块闭合盆地之内，用地西侧靠山，东侧临道路和水塘。场地内原有两户相邻的院落民居，拆除后遗留下房基和部分石墙。作为公社露营基地的配套服务中心，包含更衣、休息、餐厅、咖啡、会议、办公等功能空间。项目试图保留当代的视角，创造一种"熟悉的陌生感"，而非将视线局限在所谓的"传统"，在空间记忆、地理环境、在地建造三个层面上做出回应，探索一种包含隐喻的、在土地中自然生长的现代性。

　　新建筑被视为老宅的延续，保留老宅房基、轮廓尺寸和石墙遗迹，设置火塘、院落及"寨门"，让过去的空间与尺度随之延续在场地中；小溪接通山泉，保留场地中的老井，采集天然水资源为景观和生活所用。层层石阶时而隆起、时而下陷的起伏形态是对楼纳大尺度喀斯特地貌、地质环境的象征性重现，整个建筑犹如巨石匍匐于当地特有的喀斯特"馒头山"山脚，与楼纳的独特地景融为一体。当人从田埂间望去，所见既是大地向山林隆起的一部分，又是一个可以自由登高观景的平台，亦是一个温馨的居所。现代公共功能的置入顺应原有老宅的位置关系，以院落的方式围合，同时将两个宅基之间的空地设置为第三个内院，一侧向荫翳的自然山林敞开，当人们从开阔地带逐步进入安静的院落及屋后绿荫下的廊道，一种在公共环境下的私密感被逐渐诱发。各个房间的屋面通过平台和阶梯连接成一体，将火塘、广场、庭院、水池等地面的多样活动引向屋面，整个大冲组的山水地景尽收眼底。

　　当地人将混凝土与多种在地材料（尤其是石材）结合，形成墙角、门头、挑檐、挑台、楼梯，服务于在地生活的空间创造，并因其跨度及可塑性，极大丰富了民居的空间类型，在构造设计上沿用这些做法，并改良其工艺，发掘其塑造空间的潜力，使之为现代空间服务。餐厅使用的混凝土十字柱是石砌十字柱的改良，较大的支撑跨度为室内使用创造了灵活性，同时解放了建筑立面，使其如同一个漂浮在水上的亭榭。

摄影 / Photographer:
张广源 / Zhang Guangyuan

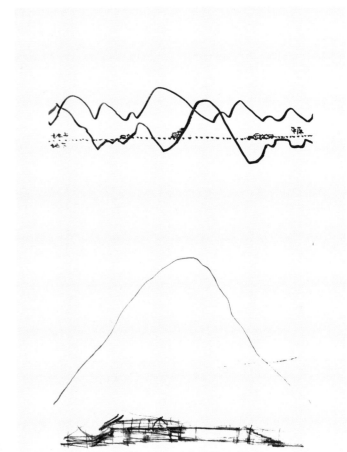

The Camping Service Center in Louna was initially designed in November 2016 and completed and put into use in October 2017. It is located in the Architect Commune of the Dachong Group, Louna Village, located in the eastern mountainous area of Xingyi City, Guizhou Province. The site is a closed basin surrounded by mountains. The west side of the site is close to a mountain, while a road and a pool appear to its east. There are two adjacent courtyard houses on the site whose foundations and some stone walls were reserved after their demolition. As the supporting service center of the commune's base camp, it includes rooms for dressing, rest, dining, meeting and a cafe, office rooms. We opted for a contemporary design that would result in a "familiar sense of strangeness". Rather than focusing on so-called tradition, we innovated at three levels: spatial memory, geographical environment and local construction, and explored a kind of modernity including metaphor and natural growth within the land.

The new building is regarded as the continuation of the preexisting house, as the building's base, contour, size and stone wall remain well preserved. The fire pits, courtyard and zhaimen (gate of a village) are arranged to represent the previous site's space and scale. The stream emerges from the mountain spring. The old well on the site is also preserve and the natural spring water collected for landscaping and service center and campsite use. The undulating stone steps rise and fall to represent the karst landforms and distinctive geology in Louna. The building is like a boulder emerging from the foot of the nearby karst mountain nicknamed the "mantou" mountain after its resemblance to steamed buns. Thus the Camping Service Center integrates with the unique landscape of Louna. When visitors approach from edge of the field, they see the changing elevation of the mountain floor to the mountain forest. They also see a platform for views and free climbing. The inclusion of modern facilities open to the public nonetheless conforms to the sitings of the original dwellings. The open space between the two homesteads is designed as the third inner courtyard, with one side open to the shaded natural mountain forest. When people gradually enter the quiet courtyards and corridors under the green shade behind the house from the open area, a sense of privacy in the public environment is gradually induced. The roofs of each wing are connected into a whole through platform and staircases. From here, visitors are rewarded with a panoramic view of the entire Dachong group landscape.

For construction, the locals mix concrete with available materials, especially stone. This material, which was novel to us, were used in living spaces such as in room corners, above door frames, on the cantilevered roof, on stairs, and on landings. Using locally inspired materials enriched the space with a vernacular feel while showcasing the timelessness and adaptive qualities of traditional constructive methods. The construction of this project follows these practices and improves the craftsmanship to make it serve modern space. The concrete cruciform columns used in the dining hall are an improvement of the local stone cruciform columns. The large supporting span created flexibility for interior use, enabling the dining hall to accommodate activities such as reception, lectures and exhibitions, while liberating the facade's design. Combined with the layout of the pool in the outer courtyard, the dining hall appears to be a pavilion floating on water.

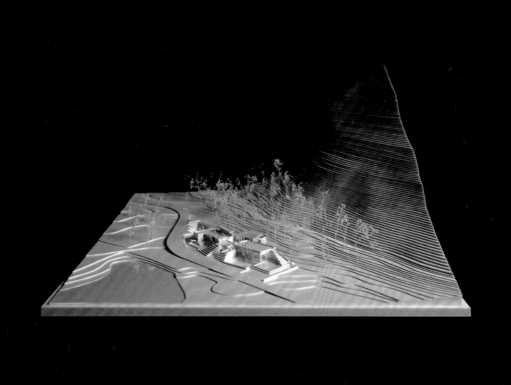

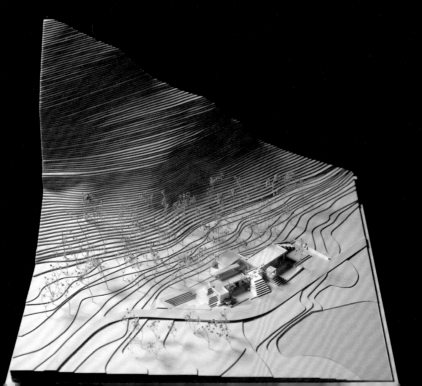

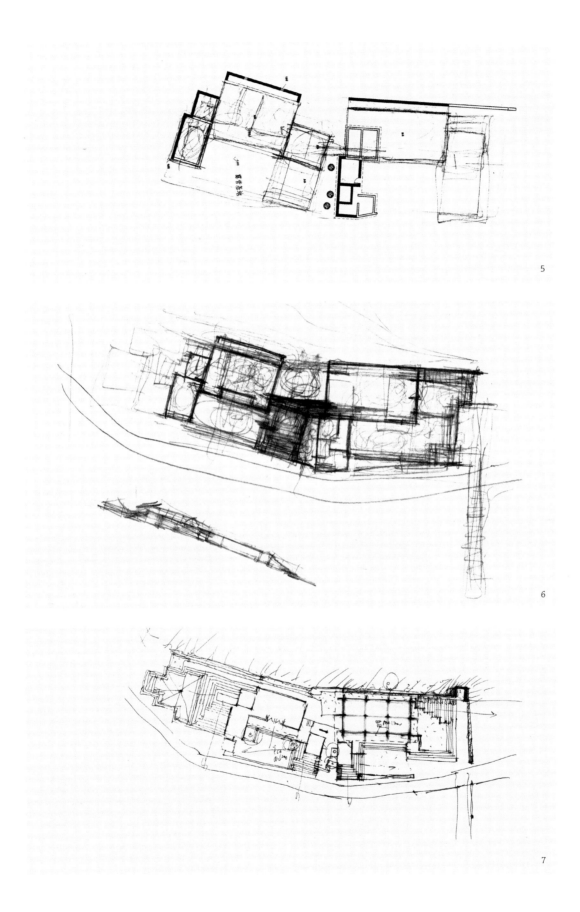

5

6

7

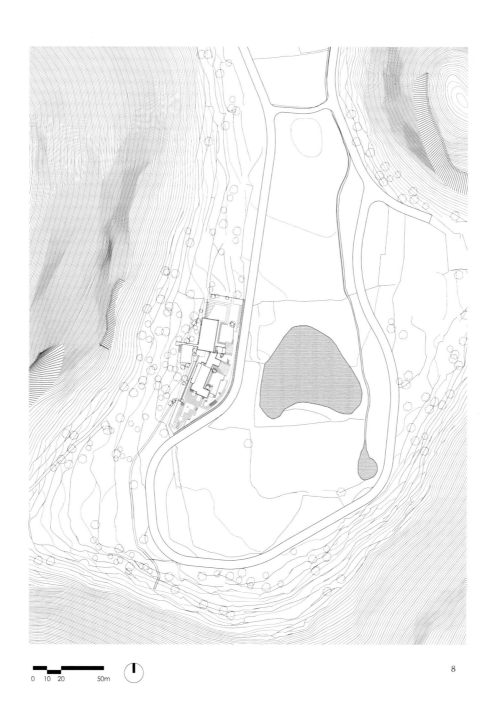

0 10 20   50m

8

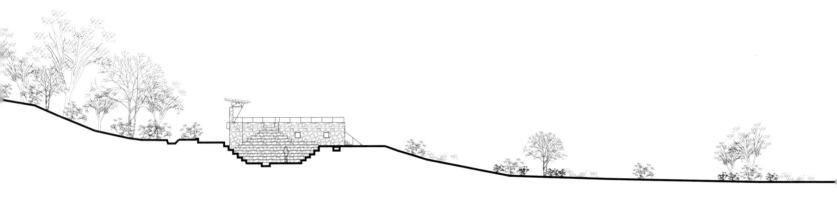

10

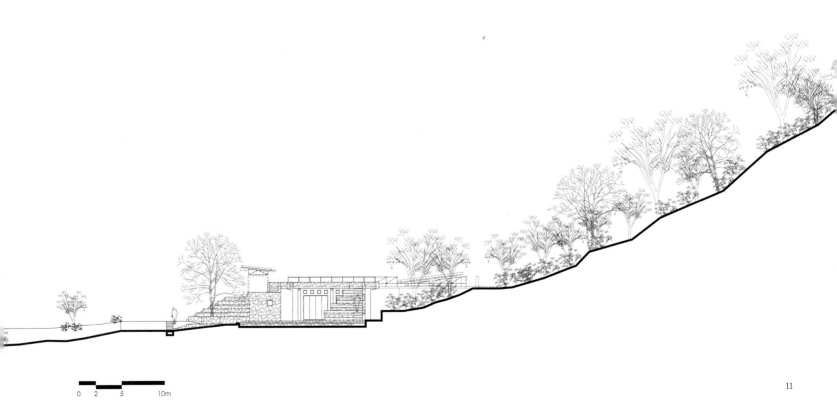

0 2 5   10m

11

12

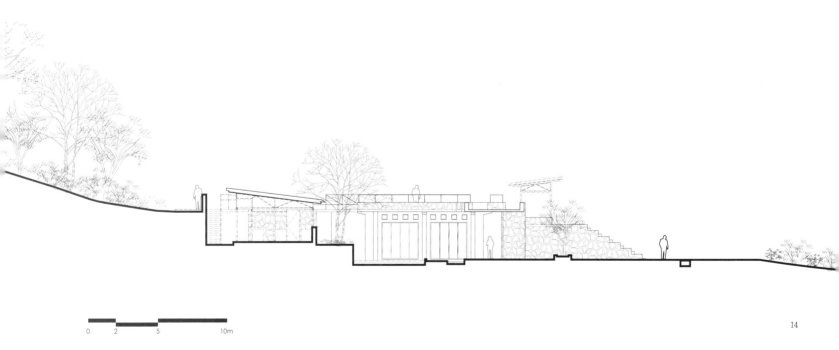

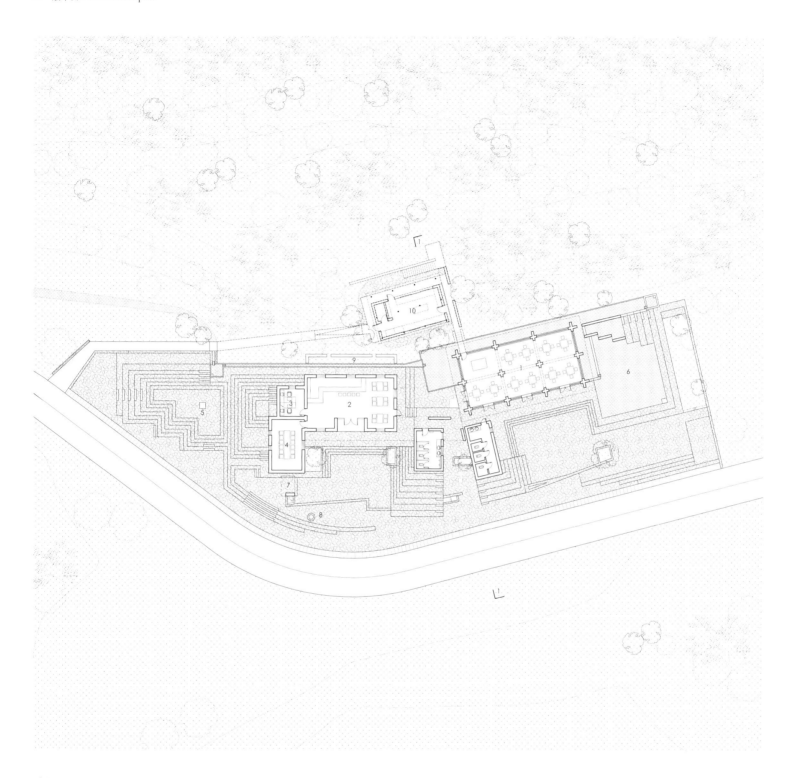

1 餐厅 / Dining hall

2 咖啡 / Cafe

3 办公室 / Office

4 会议室 / Meeting room

5 火塘 / Fire pool

6 水塘 / Pool

7 寨门 / Zhaimen

8 老井 / Old well

9 老墙 / Old stonewall

10 开放厨房（未建成）/ Open
kitchen (not completed)

0 2   5   10m   15

16

18

19

20 由内院望向外部 / View of the exterior from the courtyard
21 内院人视 / View of the courtyard

21

22 施工中的石墙 / The stone wall under construction
23 施工中的混凝土十字柱餐厅 / The dining hall with concrete cruciform columns under construction
24 屋面及墙身详图 / Details of the roof and the wall

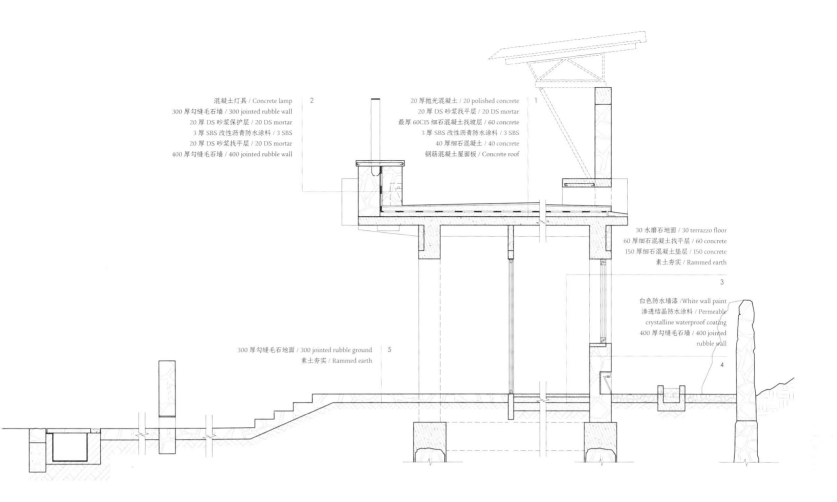

混凝土灯具 / Concrete lamp
300 厚勾缝毛石墙 / 300 jointed rubble wall
20 厚 DS 砂浆保护层 / 20 DS mortar
3 厚 SBS 改性沥青防水涂料 / 3 SBS
20 厚 DS 砂浆找平层 / 20 DS mortar
400 厚勾缝毛石墙 / 400 jointed rubble wall

2

20 厚抛光混凝土 / 20 polished concrete
20 厚 DS 砂浆找平层 / 20 DS mortar
最厚 60C15 细石混凝土找坡层 / 60 concrete
3 厚 SBS 改性沥青防水涂料 / 3 SBS
40 厚细石混凝土 / 40 concrete
钢筋混凝土屋面板 / Concrete roof

1

30 水磨石地面 / 30 terrazzo floor
60 厚细石混凝土找平层 / 60 concrete
150 厚细石混凝土垫层 / 150 concrete
素土夯实 / Rammed earth

3

白色防水墙漆 / White wall paint
渗透结晶防水涂料 / Permeable crystalline waterproof coating
400 厚勾缝毛石墙 / 400 jointed rubble wall

4

300 厚勾缝毛石地面 / 300 jointed rubble ground
素土夯实 / Rammed earth

5

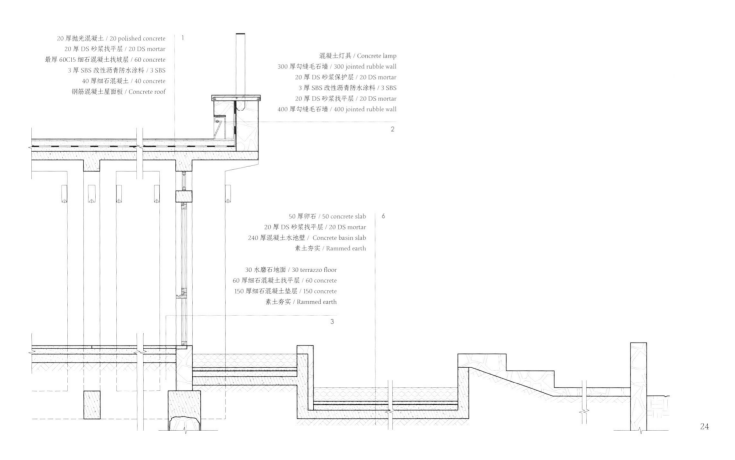

20 厚抛光混凝土 / 20 polished concrete
20 厚 DS 砂浆找平层 / 20 DS mortar
最厚 60C15 细石混凝土找坡层 / 60 concrete
3 厚 SBS 改性沥青防水涂料 / 3 SBS
40 厚细石混凝土 / 40 concrete
钢筋混凝土屋面板 / Concrete roof

1

混凝土灯具 / Concrete lamp
300 厚勾缝毛石墙 / 300 jointed rubble wall
20 厚 DS 砂浆保护层 / 20 DS mortar
3 厚 SBS 改性沥青防水涂料 / 3 SBS
20 厚 DS 砂浆找平层 / 20 DS mortar
400 厚勾缝毛石墙 / 400 jointed rubble wall

2

50 厚卵石 / 50 concrete slab
20 厚 DS 砂浆找平层 / 20 DS mortar
240 厚混凝土水池壁 / Concrete basin slab
素土夯实 / Rammed earth

6

30 水磨石地面 / 30 terrazzo floor
60 厚细石混凝土找平层 / 60 concrete
150 厚细石混凝土垫层 / 150 concrete
素土夯实 / Rammed earth

3

24

2018

"仓阁"
——首钢工舍智选假日酒店
"SILO PAVILION",
HOLIDAY INN EXPRESS
BEIJING SHOUGANG

"仓阁"——首钢工舍智选假日酒店设计始于 2015 年 12 月，2018 年 5 月竣工，位于首钢老工业区北部，原为三高炉空压机站、返焦返矿仓、低压配电室、N3-18 转运站等 4 个工业建筑，改造后成为一座特色精品酒店，同时为紧邻的北京 2022 冬奥组委办公区员工提供倒班住宿服务。"仓阁"的设计尊重工业遗存的原真性，延续历史记忆，通过新与旧的碰撞使场所蕴含的诗意和张力得以呈现，它曾经是首钢厂区生产链条上不可或缺的一个节点，今天则是城市更新的一次生动实践，并与北京 2022 冬奥会的可持续理念高度契合。

项目最大限度地保留了原来废弃和预备拆除的工业建筑及其空间、结构和外部形态特征，将新结构见缝插针地植入其中并叠加数层，以容纳未来的使用功能：下部的大跨度厂房——"仓"作为公共活动空间，上部的客房层——"阁"漂浮在厂房之上。被保留的"仓"与叠加其上的"阁"并置，形成强烈的新旧对比，"阁"在玻璃和金属材料的基础上局部使用木材等具有温暖感和生活气息的材料，使"仓阁"在人工与自然、工业与居住、历史与未来之间实现一种复杂微妙的平衡，新旧建筑相互穿插创造出一个令人兴奋的内部世界。

"仓阁"北区由原三高炉空压机站改造而成，原建筑的东、西山墙及端跨结构得以保留，吊车梁、抗风柱、柱间支撑、空压机基础等极具工业特色的构件被戏剧性地暴露在大堂公共空间中，新结构则由下至上层层缩小，屋顶天光通过透光膜均匀漫射到环形走廊，使整个客房区域充满宁静氛围；"仓阁"南区由原返焦返矿仓、低压配电室、N3-18 转运站改造而成，3 组巨大的返矿仓金属料斗与检修楼梯被完整保留在全日餐厅内部，料斗下部出料口被改造为就餐空间的空调风口与照明光源，上方料斗的内部被别出心裁地改造为酒吧廊；客房层出檐深远，形成舒展的水平视野，在阳台上凭栏远眺，可俯瞰首钢的高炉工业遗迹和远处石景山的自然风光。

设计过程中，对原建筑进行全面的结构检测，确定了"拆除、加固、保留"相结合的结构处理方案；使用粒子喷射技术对需保留的涂料外墙进行清洗，在清除污垢的同时保留了数十年形成的岁月痕迹和历史信息。

摄影 / Photographer:
陈颢、邢睿、郑旭航 / Chen Hao, Xing Rui, Zheng Xuhang

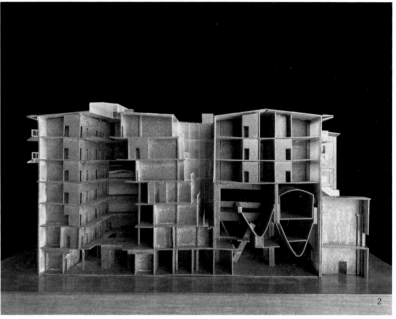

The "Silo Pavilion", Holiday Inn Express Beijing Shougang was initially designed in November 2015 and completed in July 2018. Located in the northern part of the old industrial zone of Shougang, it was transformed from four industrial buildings (the three-blast furnace air compression station, the returning coke and ore bunker, the low-voltage power distribution room, and the N3-18 transfer station) into a distinctive boutique hotel. It also provides accommodation for the staff of the 2022 Beijing Winter Olympics Office, which is adjacent to the hotel. The design respects the authenticity of the industrial remains, continues the historical memory of the old industrial zone of Shougang, and presents the harmony and tension of the building. The old and the new physically seemingly collide while integrating in function and form. The silo was once an indispensable cog in the production chain of the Shougang Plant. Today, it is a vivid example of the goal of the 2022 Beijing Winter Olympics to combine urban renewal with sustainability.

Previously abandoned and designated for demolition, these industrial buildings were now to be preserved to the utmost extent—retaining their spatial, structural, and exterior features. New structures would be constructed within them and stacked into several layers to accommodate future use. The lower space or "Silo" would be used for public space. The guest rooms would be installed in the "Pavilion", which floats above the original factory building. The juxtaposition of "Silo" and "Pavilion" forms a strong contrast between old and new. Some parts of "Pavilion" have been augmented with new components, such as metal awnings and outdoor stairs. Aside from glass and metal, materials with warmth and vitality such as wood have been partially used. The "Silo Pavilion" realizes a complex and subtle balance between artificial and nature, industry and dwelling, and the past and future. The old and new buildings interspersed here create an exciting inner world.

The north part of "Silo Pavilion" was reconstructed from the original three-blast furnace air compressor station. The east and west gables, as well as the end span structures of the original building, were preserved. The components of distinctive industrial characteristics such as the crane beam, wind-resistant column, inter-column support, and the air compressor foundation are dramatically on view in the public space of the lobby. The new structure shrinks in volume from the lower to upper floors. The skylight diffuses light evenly through the transparent membrane to the circular atrium, filling the entire guest room area with a feeling of tranquility. The south part of "Silo Pavilion" was transformed from the returning coke and ore bunker, the low-voltage power distribution room and the N3-18 transfer station. The three sets of huge metal hoppers of returning ore bunker and overhauled stairs are completely retained inside the all-day dining room. The bottom discharge hole of the hopper is transformed into the air conditioning vent and the dining area's lighting source. The interior of the upper hopper is transformed into a bar gallery in an ingenious way, that allows guests enjoy a unique space. The guest rooms have deep eaves, which allows for a panoramic view. Leaning on a balcony and looking at the distance, one can peer down at the blast furnace industrial remains of Shougang while marveling at the natural scenery of the distant Mount Shijingshan.

During the design process, architects and structural engineers worked closely together to carry out a comprehensive structural inspection of the original building, determined the structural treatment plan combined with "demolition, reinforcement and retention", and used particle jet technology to clean and preserve the exterior coating wall. While carefully removing the industrial dirt, we preserved significant traces of the site's history for decades.

4

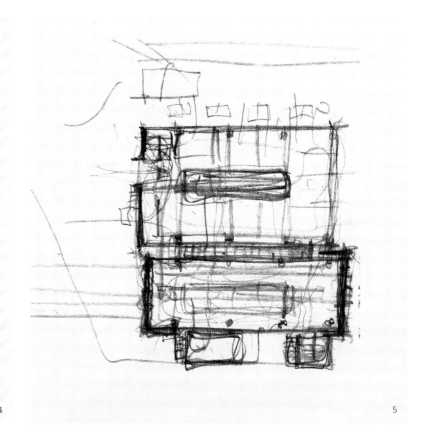

5

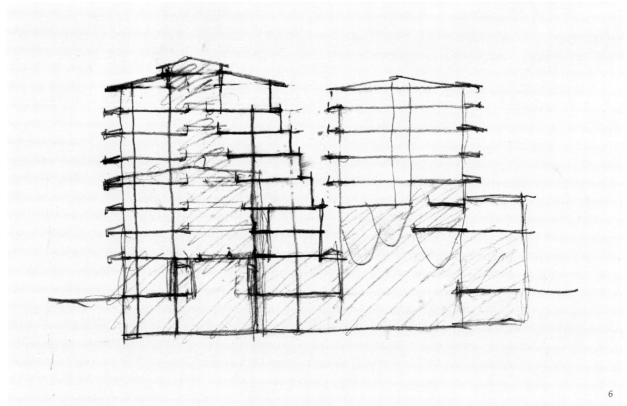

6

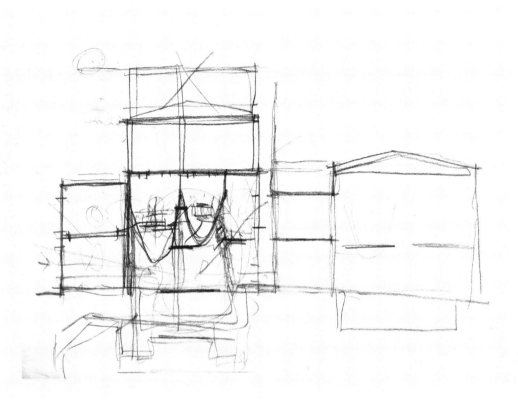

7

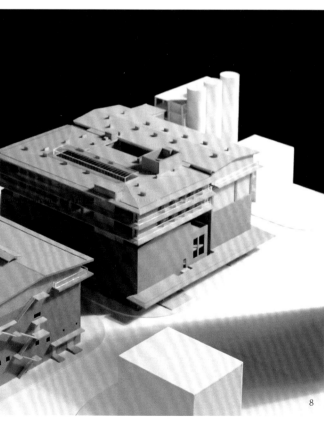

8

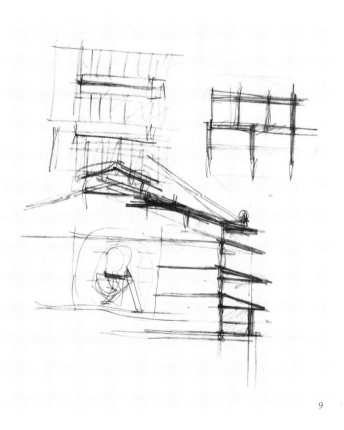

9

10

11

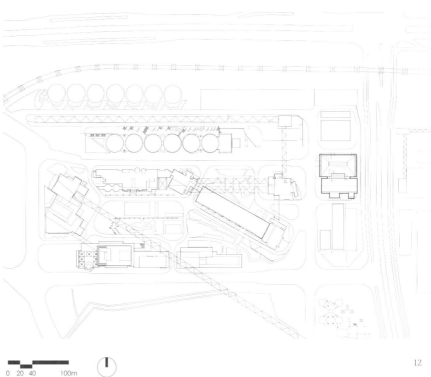

12

0 20 40   100m

13

14

1 入口门厅 / Entrance hall
2 大堂 / Lobby
3 餐厅 / Dining hall
4 厨房 / Kitchen
5 咖啡厅 / Cafe
6 健身房 / Gym
7 设备间 / Equipment
8 客房 / Guest room
9 休息厅 / Lounge

15

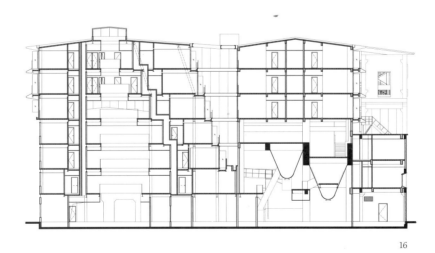

16

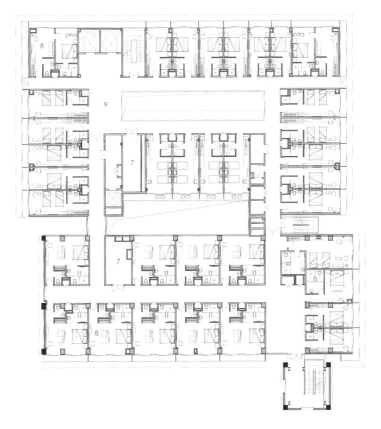

17

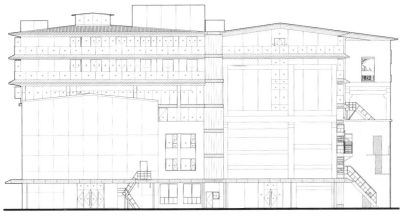

18

21

22

23

24

26-27 施工中的仓阁 / The concrete structure under construction
28 墙身详图 / Details of the wall

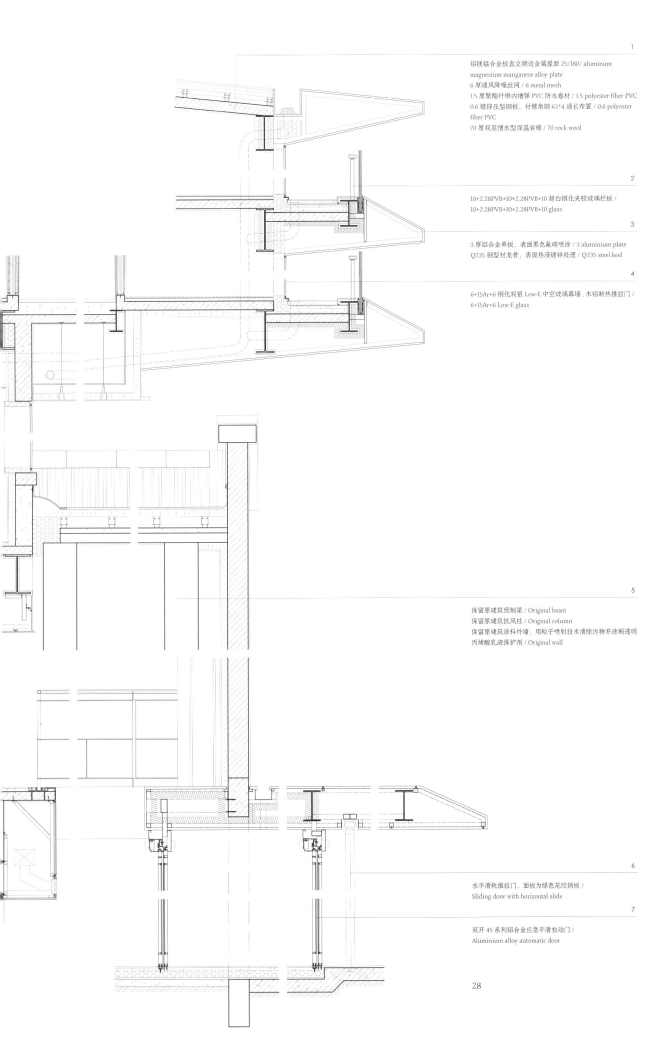

1

铝镁锰合金板直立锁边金属屋面 25/180/ aluminum
magnesium manganese alloy plate
6 厚通风降噪丝网 / 6 metal mesh
1.5 厚聚酯纤维内增强 PVC 防水卷材 / 1.5 polyester fiber PVC
0.6 镀锌压型钢板，衬檩角钢 63*4 通长布置 / 0.6 polyester
fiber PVC
70 厚双层憎水型保温岩棉 / 70 rock wool

2

10+2.28PVB+10+2.28PVB+10 超白钢化夹胶玻璃栏板 /
10+2.28PVB+10+2.28PVB+10 glass

3

3 厚铝合金单板，表面黑色氟碳喷涂 / 3 aluminium plate
Q235 钢型材龙骨，表面热浸镀锌处理 / Q235 steel keel

4

6+15Ar+6 钢化双银 Low-E 中空玻璃幕墙，木铝断热推拉门 /
6+15Ar+6 Low-E glass

5

保留原建筑预制梁 / Original beam
保留原建筑抗风柱 / Original column
保留原建筑涂料外墙，用粒子喷射技术清除污物并涂刷透明
丙烯酸乳液保护剂 / Original wall

6

水平滑轨推拉门，面板为绿色花纹钢板 /
Sliding door with horizontal slide

7

双开 45 系列铝合金应急平滑自动门 /
Aluminium alloy automatic door

28

2019

屺园
——延庆园艺小镇文创中心
MOUNTAIN GARDEN, CULTURAL CENTER OF HORTICULTURE VILLAGE IN YANQING

屺园——延庆园艺小镇文创中心设计始于 2017 年 6 月，2019 年 4 月竣工投入使用，位于北京世界园艺博览会园艺小镇用地西南，场地东北依托传统风貌的特色小镇和花田，南侧与现代化的植物馆遥望，包含为小镇服务的文创服务和特色产业体验功能，构建了一处既"开放"又"封闭"的公共空间。

建筑化整为零，由四个单坡屋顶两两组合形成的 L 形体量围合而成。一条内巷将两个体量区隔开来，半高的基座又将彼此连缀。内巷在西北、东南对角方向设置公共漫游出入口，成为园区公共游览系统的一部分，游客经由内巷上达平台，攀至屋顶，赏望远景。使用空间在西南、东北和东南角分设出入口，互不相扰：西南体量为以展厅、讲堂和图书吧为主的文创服务空间；东北体量以制作工坊、纪念品商店等特色产业体验空间为主；基座下设置管理办公室和艺术家工作室，与服务区和产业区便捷相连。各组团可分可合，会时各安其位，统一管理；会后亦可合而为一，共同运营。

开放性之余，建筑又希望能实现某种内在叙事，从而跟周边环境中那些非自然的布景式建筑或景观保持适当距离。将建筑进行两分，"筑房拟山"，完成一种特定的景观体验——单坡屋面赋予建筑以强烈的方向性和识别性，同时也完成了象征自然物的"山"与象征人工物的"房"之间的连接和互成："房"位于东北侧，"山"与"房"的基座相连，位于西南侧。相似的坡顶形式，在室内分化成两类不同特征的空间：西南侧体量趴伏，空间呈现出洞穴般的模糊性；东北侧间架清晰，暗合了传统建筑的空间特征；西南坡顶成为东北侧"房"内特殊的人造"山"景，谓之以"房"现"山"。"山""房"、檐廊、阁楼、亭台共同构成一个立体的观游系统，故名"屺园"。

"山"的部分采用全现浇清水混凝土墙体承托斜屋面，强化"山穴"的空间特征；"房"的部分采用钢筋混凝土框架与木屋架复合的结构体系，象征传统的木作屋架。"房"中的木屋架采用正交木桁架，等截面的木杆件通过预制的多向钢构件连接，形成既现代高效又与传统榫卯木作有所关联的复合式节点。

摄影 / Photographer:
张广源、李维纳、李兴钢、姜汶林 / Zhang Guangyuan, Li Weina, Li Xinggang, Jiang Wenlin

The Mountain Garden, Cultural Center of Horticulture Village in Yanqing was initially designed in May 2017 and put into use in March 2019. It is located in the southwest grounds of the International Horticultural Exhibition 2019 in Yanqing, Beijing. The site is adjacent to the traditional model town and flower field in the northeast, and parallel to the modern Botanical Hall in the south. Aiming to provide both open and closed public spaces, it houses a site for cultural and creative activities while showcasing a traditional industry, which are the goals of Expo.

The building is divided into two L-shaped parts formed by a combination of two single slope roofs. The two volumes are separated by an inner alley but connected at the base at the half floor height. Public entrances and exits are set up in the northwest and southeast corner of the inner alley, becoming part of the public route of the town park. Tourists can reach a platform through the inner alley, climb to the upper floor and the roof, and enjoy looking at the landscape in distance. The independent entrances and exits for functional spaces are located in the southwest, northeast and southeast corner, respectively. The southwest area contains cultural and creative spaces such as a gallery, a lecture hall, and a book bar. The northeast area features commercial spaces for workshops and shops. Public rooms as management office and artist studios are underneath the base, where they are conveniently connected to the service and commercial areas. Each group of rooms can be both separable and compatible. The groups can be installed and managed in a unified way during the Expo and integrated and operated together after the event.

In addition to being open, the building attempts to express an internal narrative so as to keep a proper distance from the less organic and photogenic buildings and landscaping surrounding it. The single sloping roof gives the building a clear focus and identity, which is the concept of "building houses and imitating mountains". The "mountain" symbolizes natural objects and the "house" symbolizes artificial objects that are connected and inter-related. The "house" located in the northeast, while the "mountain" located in the southwest merges with the "house's" base which is a split-level platform. Similar sloping roofs define two kinds of interior spaces with different characteristics. The southwest quadrant is low to the ground, evoking a cave. The structure in the northeast is more defined, suggesting the traditional form of a house. Looking from the northeastern "house", the slope of the roof in the southwest appears to be a "mountain". The "mountain", the "house", the colonnade, the eave porch and attic on the upper floor, and the pavilion on the "mountain" roof form a three-dimensional sightseeing system, or what we called "Mountain Garden".

The "mountain" is made from a cast pale concrete wall that supports the sloped roof to broaden the spatial characteristics of the "mountain cave". The "house" portion is made of a composite structural system with a reinforced concrete frame and a wood-framed roof, which is inspired by traditional wooden roof design. The wood-framed "house" uses an orthogonal wooden truss. The wood components with the same cross-section are connected by prefabricated, multi-directional steel components. These components form a modern, efficient composite joint akin to traditional mortise and tenon woodworking methods.

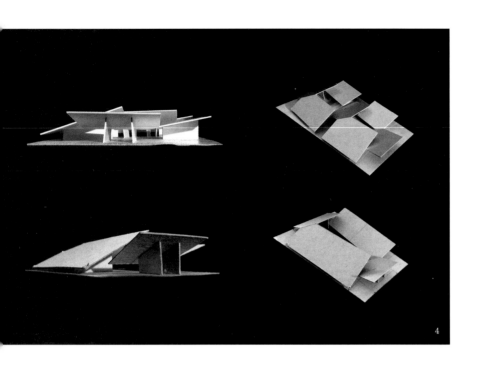

0 10 20   50m

9

11

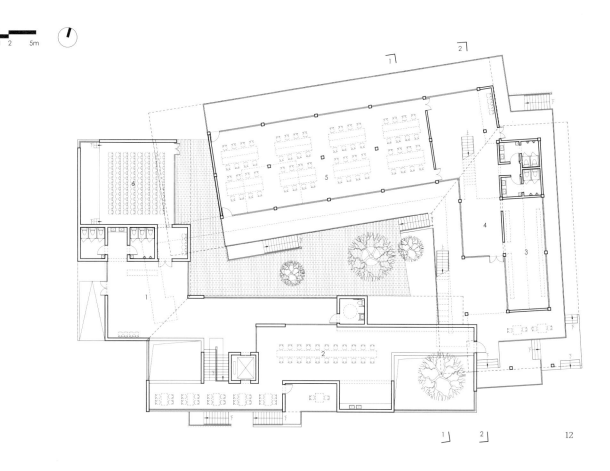

0 1 2 5m

12

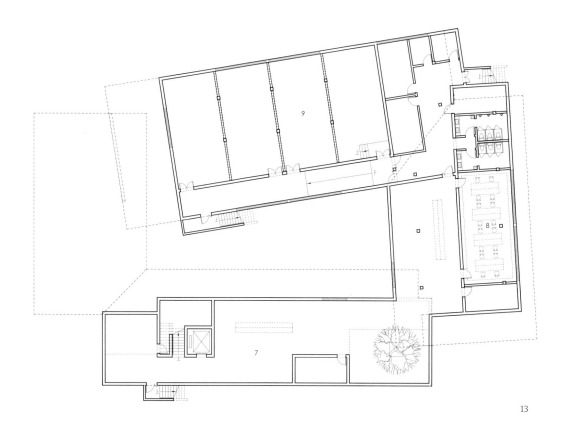

13

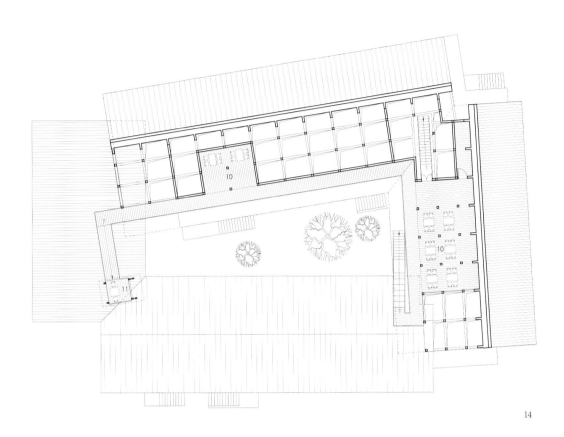

14

1 展区门厅 / Entrance hall of the gallery

2 图书吧 / Book bar

3 纪念品商店 / Shop

4 文创区门厅 / Entrance hall of the culture creation area

5 制作工坊 / Workshop

6 讲堂 / Lecture hall

7 展厅 / Gallery

8 管理办公室 / Office

9 艺术家工作室 / Artist studio

10 观景平台 / Platform under wooden frame

11 观景亭 / Pavilion

16

17 剖面 1-1/ Section 1-1
18 木屋架夹层西望 / View from the platform under the wooden frame to the west
19 剖面 2-2/ Section 2-2
20 从图书吧望向展厅 / View from the book bar to the exhibition hall

0 1 2 5m

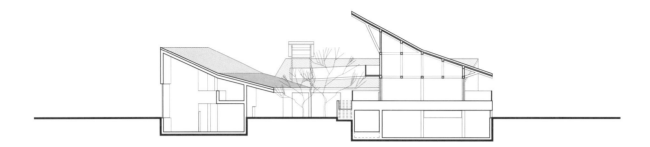

17

18

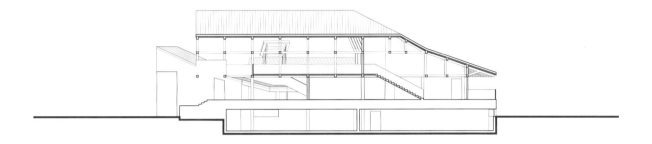

21

22

23

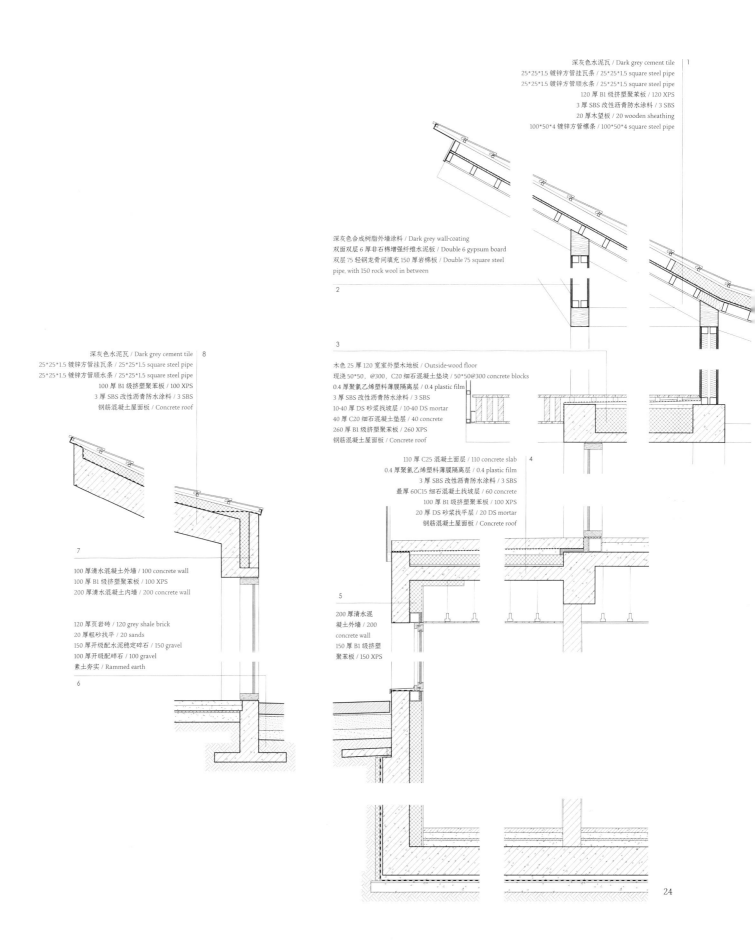

深灰色水泥瓦 / Dark grey cement tile | 1
25*25*1.5 镀锌方管挂瓦条 / 25*25*1.5 square steel pipe
25*25*1.5 镀锌方管顺水条 / 25*25*1.5 square steel pipe
120 厚 B1 级挤塑聚苯板 / 120 XPS
3 厚 SBS 改性沥青防水涂料 / 3 SBS
20 厚木望板 / 20 wooden sheathing
100*50*4 镀锌方管檩条 / 100*50*4 square steel pipe

深灰色合成树脂外墙涂料 / Dark grey wall-coating
双面双层 6 厚非石棉增强纤维水泥板 / Double 6 gypsum board
双层 75 轻钢龙骨间填充 150 厚岩棉板 / Double 75 square steel
pipe, with 150 rock wool in between

2

3

深灰色水泥瓦 / Dark grey cement tile | 8
25*25*1.5 镀锌方管挂瓦条 / 25*25*1.5 square steel pipe
25*25*1.5 镀锌方管顺水条 / 25*25*1.5 square steel pipe
100 厚 B1 级挤塑聚苯板 / 100 XPS
3 厚 SBS 改性沥青防水涂料 / 3 SBS
钢筋混凝土屋面板 / Concrete roof

木色 25 厚 120 宽室外塑木地板 / Outside-wood floor
现浇 50*50，@300，C20 细石混凝土垫块 / 50*50@300 concrete blocks
0.4 厚聚氯乙烯塑料薄膜隔离层 / 0.4 plastic film
3 厚 SBS 改性沥青防水涂料 / 3 SBS
10-40 厚 DS 砂浆找坡层 / 10-40 DS mortar
40 厚 C20 细石混凝土垫层 / 40 concrete
260 厚 B1 级挤塑聚苯板 / 260 XPS
钢筋混凝土屋面板 / Concrete roof

110 厚 C25 混凝土面层 / 110 concrete slab | 4
0.4 厚聚氯乙烯塑料薄膜隔离层 / 0.4 plastic film
3 厚 SBS 改性沥青防水涂料 / 3 SBS
最厚 60C15 细石混凝土找坡层 / 60 concrete
100 厚 B1 级挤塑聚苯板 / 100 XPS
20 厚 DS 砂浆找平层 / 20 DS mortar
钢筋混凝土屋面板 / Concrete roof

7

100 厚清水混凝土外墙 / 100 concrete wall
100 厚 B1 级挤塑聚苯板 / 100 XPS
200 厚清水混凝土内墙 / 200 concrete wall

120 厚页岩砖 / 120 grey shale brick
20 厚粗砂找平 / 20 sands
150 厚开级配水泥稳定碎石 / 150 gravel
100 厚开级配碎石 / 100 gravel
素土夯实 / Rammed earth

5

200 厚清水混
凝土外墙 / 200
concrete wall
150 厚 B1 级挤塑
聚苯板 / 150 XPS

6

2019

南京安品园舍
ANPIN GARDEN HOUSES
IN NANJING

南京安品园舍设计始于 2014 年 3 月，2019 年 8 月基本竣工，位于南京老城南历史城区的安品街，是一个在喧闹的"城市森林"中间由独户住宅组成的居住街区，周边被 8—15 层的住宅所包围，在南北两侧还各有一栋高层住宅，对用地内的住宅形成极为不利的视线干扰。安品园舍是在历史街区进行的低层高密度和水平延展的城市新"聚落"营造，是"宅院合一"空间模式的当代实验，是努力将"日常诗意"与"都市胜景"的体验带入更大规模人群生活中的"理想居所"。

总体布局上延续安品街片区原有"八爪金龙"的历史街巷格局，形成"街—巷—院—井"的传统街区空间组合，构成了整体统一而丰富多变的城市肌理，围合式院落沿南北主街和内环小巷错落布局，减弱周边环境对住宅私密性的不利影响。

四类基本户型——"合园""筱园""中园""岢园"，以不同方式组合构成了整个街区。"合园"在宅地尺度上接近苏州小园——残粒园，借鉴其环绕中心拾级而上的游园模式组织空间，并将宅园并置的空间模式转化为宅园合一、游居一体的内向生活空间，"合园"不断抬升的地平标高正是对"残粒园"拾级登高的抽象转化，而屋顶制高点的观景亭也正是对残粒园"栝仓亭"登高远望的意象转化。"岢园"位于社区主入口，取曾存于安品街的"可园"之谐音，又意为"可山可池"；东西两翼以环绕主庭院的廊道相连，形成有高差的、连续的漫游路径，将各功能空间串联，部分节点放大变形为亭、台、楼、阁，获得了一种在剖面与空间上的"曲折有致"；纵墙区分了主院、廊子与边院，园内的"自然物"和边界的"园墙"清晰区分：前者柔软、自然、具有生长性，后者则以抽象的几何化方式对前者进行烘托、悬置，形成对比，自然物在人工界面之中，仿佛盆景、画作；游者穿梭于"园墙"界面内外，入画出画，获得丰富的空间知觉体验。漫游路径的终点是屋顶的观景廊亭，登高观望整个社区连绵起伏的坡顶屋面以及背景中的城市景观，获得深远不尽的平远图景。

摄影 / Photographer:
LSD、刘振 / LSD, Liu Zhen

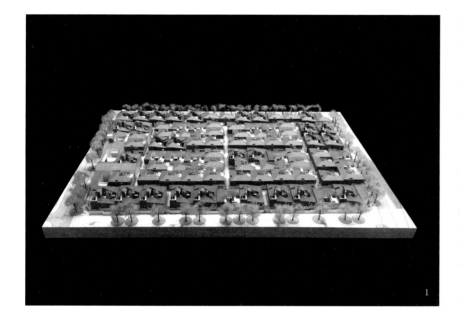

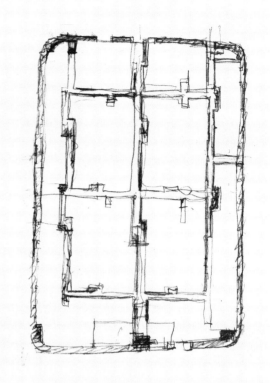

The Anpin Garden Houses in Nanjing were initially designed in March 2014 and completed in August 2019. They are located on Anpin Street in the southern historical section of the old city of Nanjing. As a residential block composed of single family houses in the middle of a noisy "urban forest", they are surrounded by 8—15 story houses. Two high-rise residential buildings on either side of the houses to the north and south can disrupt the lives of Anpin Garden House residents. Anpin Garden Houses is a new low-rise, high-density urban "settlement" extended horizontally in the historical district. As a contemporary experiment of the "house-yard integration" spatial model, it strives to bring the experience of "daily poetics" and "urban scenery" into the "ideal residence" to a larger population.

The original "eight claw golden dragon" street pattern of Anpin Street area is preserved in the general layout. This traditional street pattern is a spatial combination deemed the space composition of "street-lane-yard-patio", which creates a rich, unified urban atmosphere. Enclosed courtyards are scattered along the main north and south streets and inner ring alleys, reducing the intrusive impact of the surrounding environment on the privacy of the houses.

Four basic garden house types—"Heyuan Garden", "Xiaoyuan Garden", "Zhongyuan Garden" and "Keyuan Garden" occupy the block in different ways. The homestead scale of "Heyuan Garden" is close to that of Suzhou small garden— Remnant Grain Garden. The spatial organization of Heyuan Garden refers to the model of Remnant Grain Garden where visitors wander upwards around the center. The juxtaposition of house and garden draws the living space inward to integrate the house and garden. The rising elevation of "Heyuan Garden" is the abstract transformation of Remnant Grain Garden. The highest point of the pavilion is the viewing platform, which refers to the image of Guacang Pavilion in Remnant Grain Garden. Keyuan Garden is located at the main entrance of the community. The name of the garden is a homonym for "keyuan", which was at one time stored on Anpin Street. It could also refer to generation of mountains and waters. The east and west wings are connected by corridors around the main courtyard, which provides paths for continuous wandering through elevation changes. The functional spaces are connected along the path and interrupted by nodes that widen into pavilions, platforms, and buildings. As the path twists and turns, beautiful scenery emerges. The vertical wall separates the main courtyard, corridors and side courtyards. The "natural objects" in the garden and the "garden wall" in the boundary are clearly distinguished. The "natural objects" are soft, natural, and alive, while the "garden wall" provides an abstract and angular contrast for the natural objects. The natural objects viewed through an artificial interface are like bonsai and painting. Tourists shuttle in and out of the "garden wall" interface, moving in and moving out, and receive a complete experience from different perspectives. At the end of the wandering path sits the observation gallery in a rooftop pavilion. Here, visitors climb the community's rolling, sloped roof and gaze upon the distant urban landscape.

0 5 10   30m

5

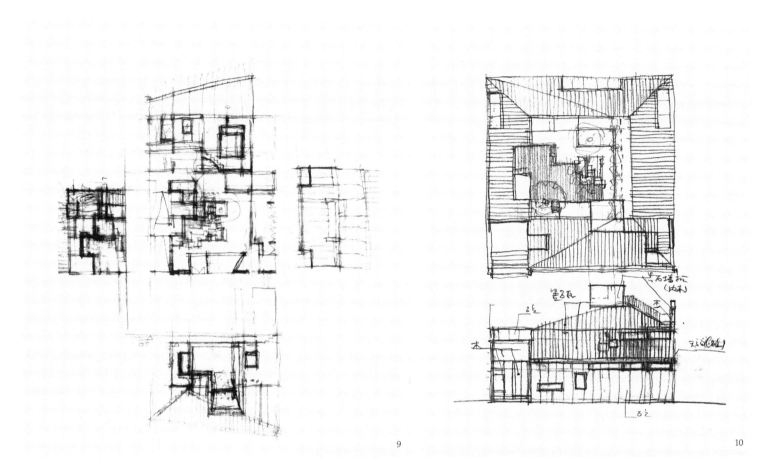

9

10

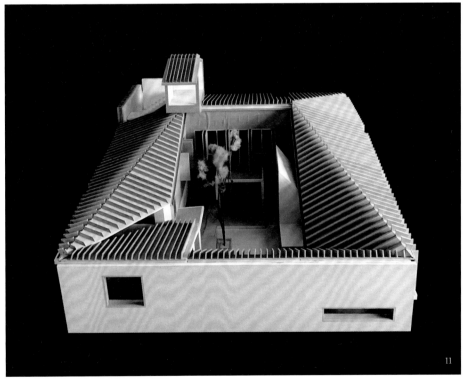

11

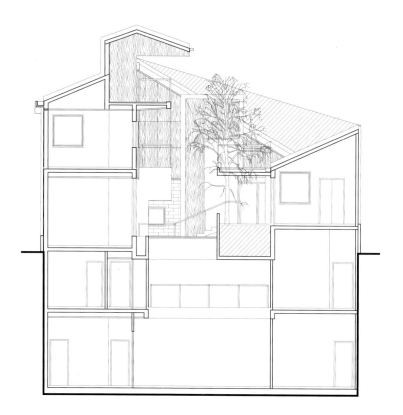

1 玄关 / Foyer
2 客厅 / Living room
3 卫生间 / Lavatory
4 中厨 / Chinese kitchen
5 西厨 / West kitchen
6 餐厅 / Dining hall
7 书廊 / Reading corridor
8 书房 / Study room
9 次卧 / Bedroom
10 主卧 / Main bedroom
11 平台 / Platform

13

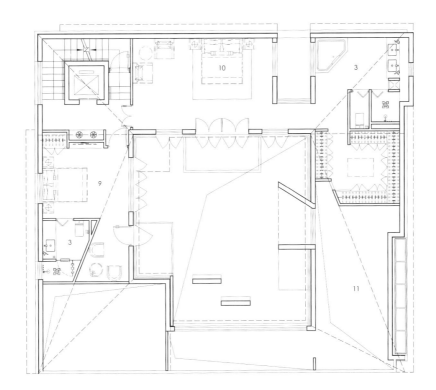

14

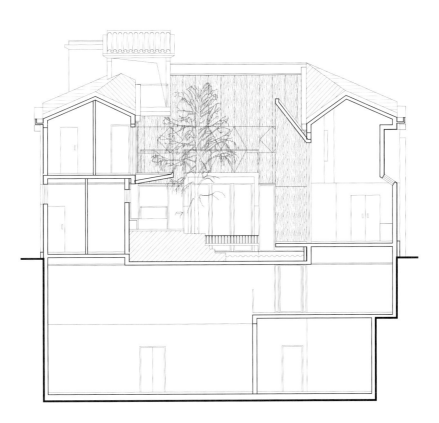

15

16

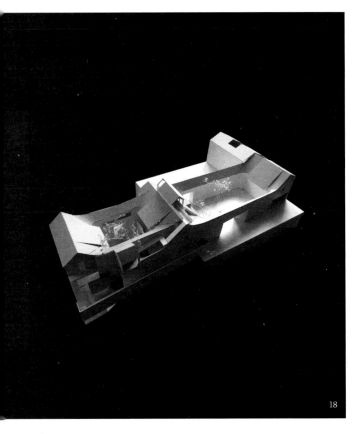

18

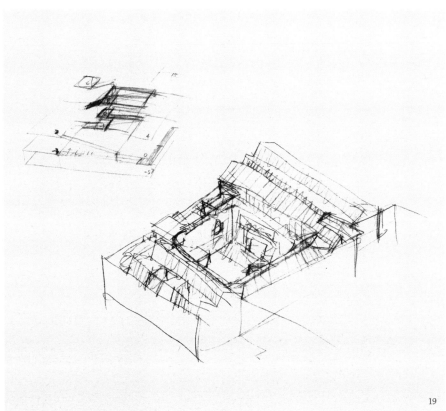

19

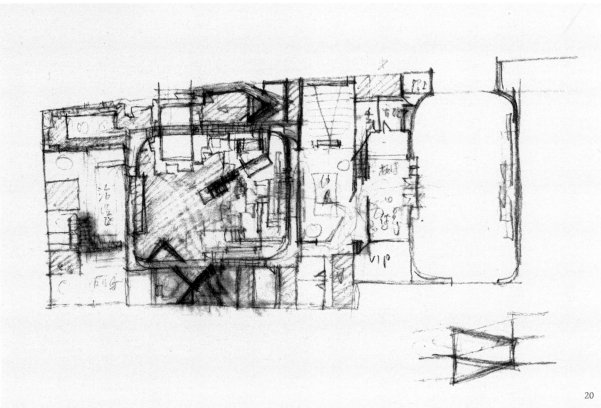

20

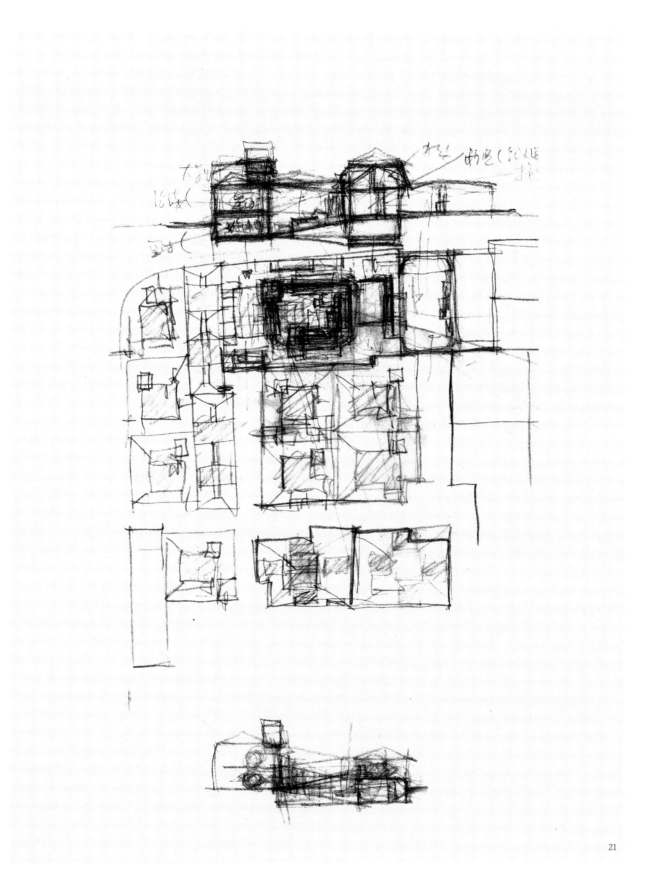

21

22 岢园一层平面 / The 1st floor plan of the Keyuan Garden
23 岢园二层平面 / The 2nd floor plan of the Keyuan Garden
24 岢园剖面模型（1/50）/ Section model of the Keyuan Garden (1/50)
25 岢园剖面 1-1/ Section 1-1 of the Keyuan Garden

0 1 2   5m

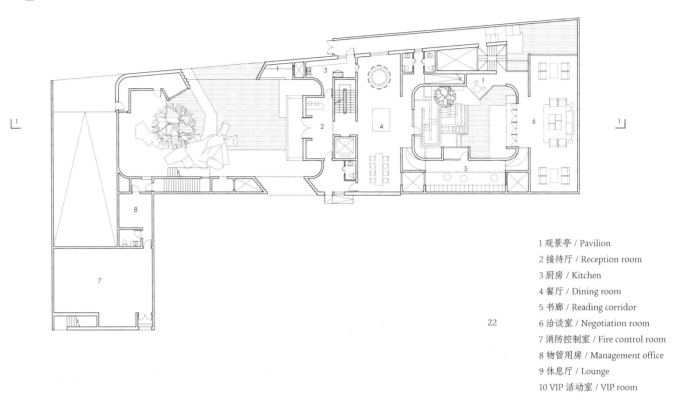

1 观景亭 / Pavilion
2 接待厅 / Reception room
3 厨房 / Kitchen
4 餐厅 / Dining room
5 书廊 / Reading corridor
6 洽谈室 / Negotiation room
7 消防控制室 / Fire control room
8 物管用房 / Management office
9 休息厅 / Lounge
10 VIP 活动室 / VIP room

22

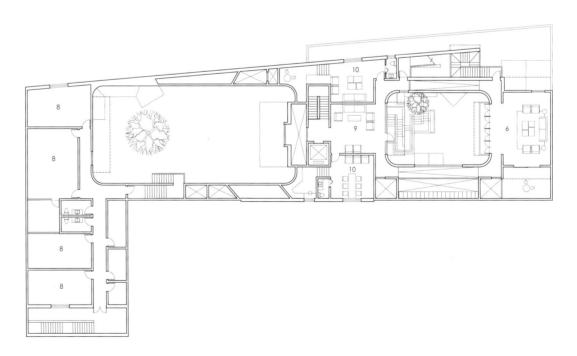

23

24

0 1 2   5m

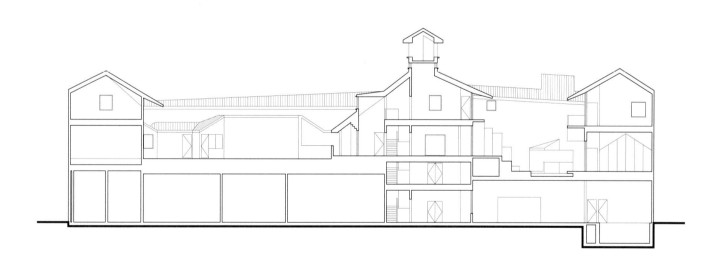

25

2019

璧园
BIYUAN GARDEN

璧园位于北京市海淀区一个普通小区内，是一栋普通住宅楼地下一层的一个下沉庭院。2018 年，受友人之邀，为其自宅作园，结合宅内书房，供日常休憩、会客之用。

庭院位于宅南，东西仅十米，南北不足五米。东、西、南三面高墙合围，宅院之间则为楼体承重墙，仅余三处洞口相通，宅中由东至西划为三间：卧室、库房、书房各得一洞口，日光掠过高墙入院已减至七成，至室内则不足三成。于是拓宅入院，将院北三分之一划为室内，南向遍设落地门窗，使园景如长卷映照入室，西接书房、北接库房，使宅中明暗调和，是为茶室。出茶室入园，有木栈台环绕，略高于庭院，可通行亦可闲坐休憩，栈台于东南侧外扩，斜挑入院中，成四方状，将庭院虚界为东北、东南、西侧三部分，于其上立柱架檐，是为小轩。又紧邻南壁加筑一道曲墙，上设披檐，从院中界出一条幽长狭缝连缀东西，是为曲廊。曲墙有树叶、梅花、葫芦状空窗可窥院景，中部开圆洞为门，可由廊内与茶室相望，并设木桥相通。廊内青砖墁地，行至西侧角落，地面逐级抬升，于墙角扩大成台，曲墙压低，弯转向北，廊檐亦随之折转，是为角亭。至此，一房一室，一桥一廊，一亭一轩，高下错落，相望生趣。园西曲墙环围，正对书房，又位于小轩视廊尽端，是对景之要处，在此设一池一峰，并栽细竹映衬；园东轩外花草繁荫，供观赏，亦作菜田；于东南置矮峰，可观于茶室。园中以青石板铺地，栈台、木桥、小轩皆似漂浮其上，又置零星卧石点缀，观游其间，若置身溪涧。曲廊中段略向外拓，廊内上衔天光，下种修竹，自圆洞外观之，光拂南壁、竹影斑驳，皆压合入框成画，曲墙若浅雕浮出南壁。盈尺之地造园，亦如砖雕、木雕，于一壁之中而造天地，又由壁上圆孔，取玉璧之"璧"义而避"面壁"之意，是谓"璧园"。

造园于盈尺，构筑皆需轻巧。廊亭均以竹钢为柱，钢板为顶，覆薄石板于其上，又在檐口上翻钢板，裁为"方齿"状，作滴水之用。曲墙长十米，高二米有余，从窄院"窃"得一廊已使园中局促，墙体更需纤薄。故在墙中藏设极细钢柱，砌八厘米预制空心砌块于其间。砌块有方孔，竖向贯通，配筋灌以砂浆，结成暗"柱"，而后挂网抹灰，厚仅十厘米，最后通体漆白，略泛天光，轻柔如纸。园林峰石大而厚重，不宜于此地下室上的狭小庭院搬运营建，且佳石难觅，非机缘不可寻。于是，随园主凑得小石十余块，以圆锯故作切割，弃其糟粕，以抛光切面相拼，聚而成峰。园中卧石、矮峰皆按此法。石乃天成，人精工而聚之，可谓人作天工之互成。

摄影 / Photographers:
张广源、杨玲、李兴钢、侯新觉 / Zhang Guangyuan, Yang Ling, Li Xinggang, Hou Xinjue

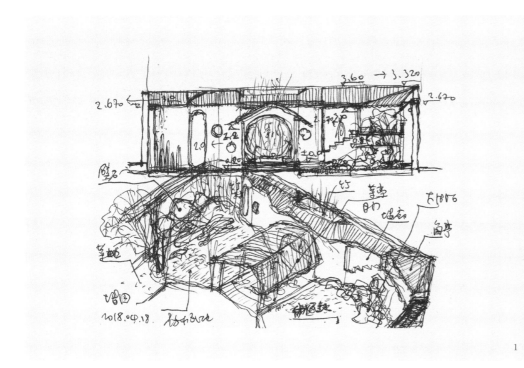

Biyuan Garden is located in the sunken basement courtyard of an ordinary residential building in Haidian District, Beijing, only ten meters wide from east to west and less than five meters deep from north to south. I decided to expand the apartment to the south, transforming one-third of the courtyard into indoor space. I added floor-to-ceiling doors and windows on the south side. Seen from the interior, the garden view looks like a long scroll painting. The expanded space is connected to the study on the west and used as a tea room.

The boardwalk is set along the doors and windows, slightly higher than the floor of the courtyard where you can continue on or rest, and extends diagonally into the courtyard on the southeast side, forming a square shaped veranda (*xiaoxuan*). A curved wall is built next to the south wall and keeps a long and narrow slit with it. The wall with a cornice set on the roof connects the east wall to the west wall and forms a curved corridor (*qulang*) in the courtyard. The curved wall's three leaf, plum blossom, and gourd-shaped portholes allow views of the garden's interior. A circular hole in the middle of the wall is also an entrance and connected with the tea room by a wooden bridge. The path is paved with grey bricks, and one can find that the ground is gradually raised and expanded into a platform at the corner. The eave of the corridor also curves around, forming a corner pavilion (*jiaoting*). Here, the study, the tea room, the bridge, the corridor, the pavilion, and the veranda, are scattered from place to place, facing each other playfully. The western part of the garden is at the end of the viewing path from the veranda. As the main point of scenery, a pool and a stone peak are located here. The eastern part of the garden is rich with flowers and plants, and also used as a vegetable field. The garden paths are paved with bluestone slabs, the boardwalk, the wooden bridge, and the tea pavilion all seem to be floating above the ground, sporadically decorated with stones. People visiting and touring the garden feel as if they were in a mountain stream. The middle section of the curved corridor extends slightly outwards, forming a small patio with a skylight from the top and bamboo growing from below. The sunlight shining through the round skylight and the mottled bamboo form and frame shadows on the wall. The curved wall looks like a shallow carving floating out from the south wall. Building a garden from previously unlandscaped land is like carving out a small world from brick or wood. Because the garden is surrounded by walls, and the circular holes opened on the middle of the main wall, which is like the shape of ancient Jade "Bi". "Bi" is a homonym for "wall" in ancient Chinese. So the tiny garden is named Biyuan Garden.

Because of the limited existing area, the structures on the Biyuan Garden needed to be lightweight and handy. The corridor and pavilion are made of pressure treated bamboo fiber. The roof is a steel plate covered with thin bluestone slabs. Steel plates on the cornice are bent into a "square teeth" shape that helps divert rainwater. The curved wall is ten meters long and more than two meters high. To give the gallery walls support, we hid thin steel columns behind the walls. Between the columns are prefabricated hollow blocks that are only eight centimeters wide. After being plastering, the wall is still only ten centimeters thick. When illuminated under the skylight, it looks as soft and thin as paper.

The large and heavy garden stones were not easy to relocation. Finding appropriately sized and quality stone for the narrow courtyard was also difficult. Thus, the owner and I gathered more than ten pieces of small stone, shaped them with a circular saw, discarded the remnants, polished the selected pieces, and joined the stone to form a "peak". We used this production method to form laid stone and other short "peaks" in the garden. The stones were made by nature and finished and gathered together by man, again showing the inter-creation between the man-made and the nature-work.

3 从一层阳台望向园子 / View from the balcony on the ground floor to the garden
4 从茶室望向园中 / View from the tea room to the garden

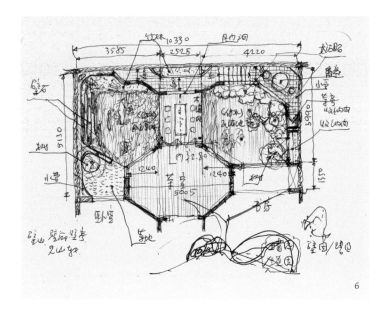

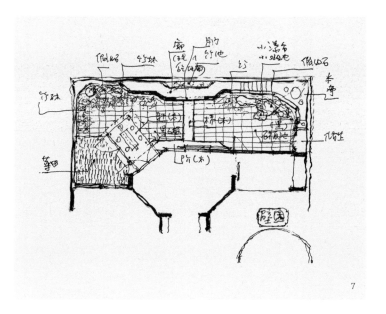

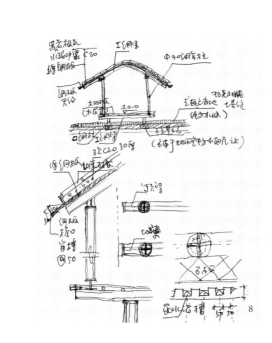

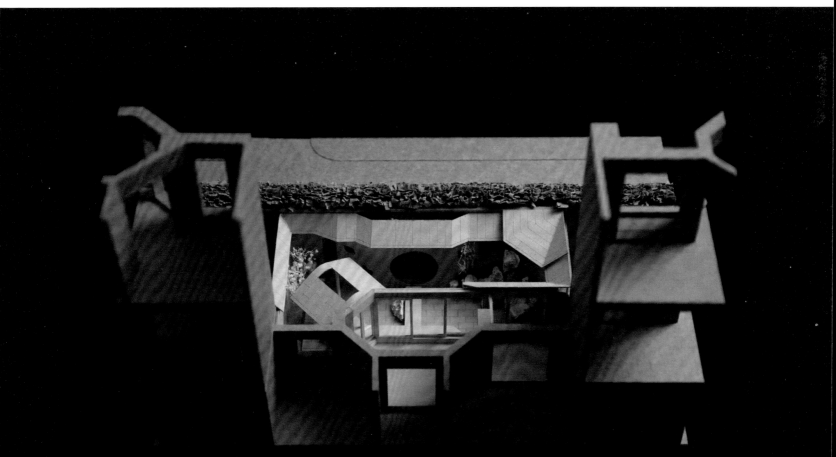

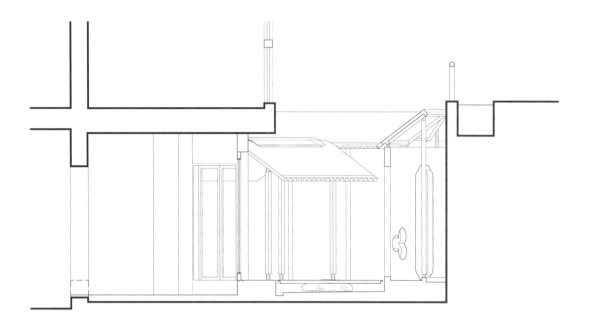

11

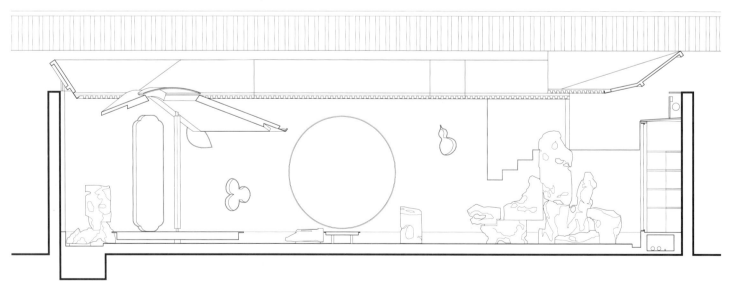

12

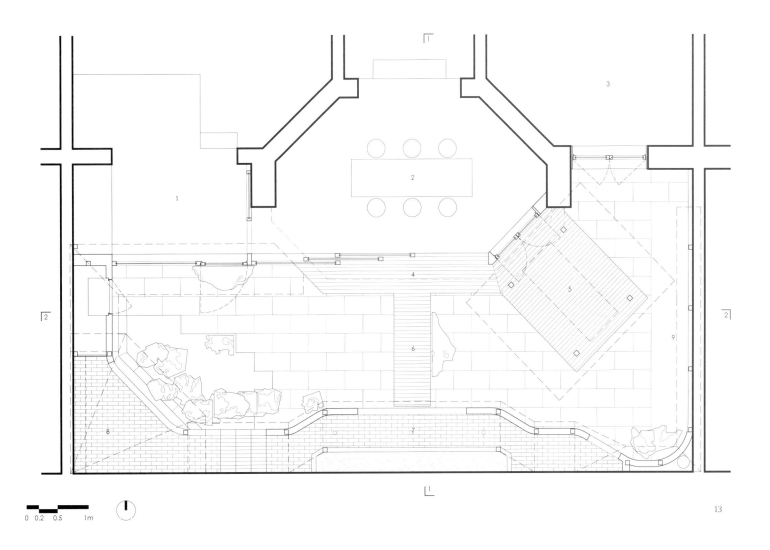

1 书房 /Study room

2 茶室 / Tea room

3 卧室 /Bedroom

4 栈台 / Boardwalk

5 小轩 /Veranda

6 桥 / Bridge

7 曲廊 / Curved corridor

8 角亭 / Corner pavilion

9 菜田 / Vegetable field

2019.06.04 壁固登山机绘
太阳石刻组合
"人之+自然"="登山" 26

27

28 施工中的竹钢＋钢结构 / The glued bamboo-steel and steel structure under construction
29 施工中的预制空心砌块曲墙 / The curved prefabricated hollow block wall under construction
30 总体轴测 / The axonometric drawing

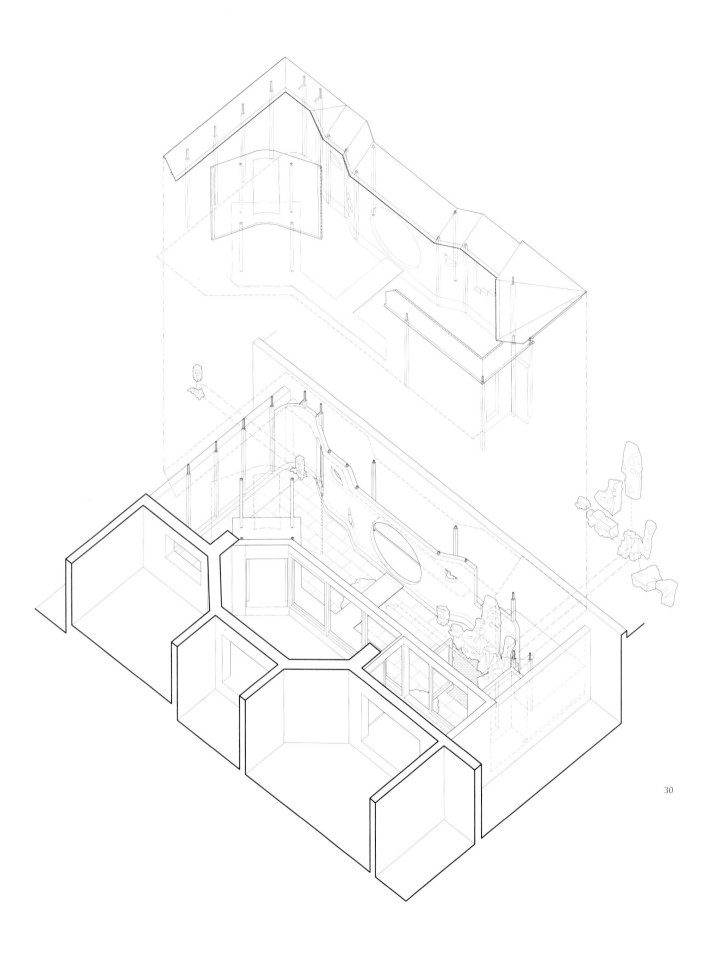

30

2020

玉环博物馆和图书馆
YUHUAN MUSEUM & LIBRARY

　　玉环博物馆和图书馆设计始于 2013 年 3 月，预计 2020 年竣工投入使用，位于浙江省台州市的海岛市——玉环市的填海新城开发区，包括图书馆、博物馆和两馆之间的公共景观空间。玉环老城的坎门一带，空间极具特色，以隧道—山—港—湾—海峡—对景岛的元素，构成了极具本地渔港特色的空间形态。在新区两馆的设计及其所在城市公园近乎"荒芜"的环境中，移植了坎门渔港的空间形态，并将两组建筑结合其间的广场和水景共同设计形成"山水之势"，在形成两馆独立使用流线的同时，营造了具有空间层次的整体环境空间。

　　博物馆和图书馆均为由曲面屋顶建筑单元进行水平和垂直方向的组合、叠加及围合而构成的现代"渔村聚落"，两组建筑群放在巨大的石砌基座之上，分别经由长长的坡道或台阶抵达，人们"穿山入港"——穿越隐藏着建筑入口的"隧道"式架空空间，进入门厅或其后半围合的内庭，建筑的空间单元之间相互咬合，串联或叠落成不同的使用空间。两组建筑的内庭相对遥望，并由长长的景观水面联通起来，相互形成彼此的对景，加上其间的"渔亭""码头""堤坝"、台阶、树林等景观元素，共同构成玉环老城的"渔港"意象，并因借远处的自然山水，合之以为"胜景"。

　　项目进一步发展了对建筑基本单元及其组合以生成建筑和组群空间的研究，由一种反曲面的混凝土悬索结构和一种类似形式的大跨度鱼腹梁结构作为基本结构和功能、空间、形式单元，在水平和垂直两个方向被反复组合、变异、连接与围合，形成独具特色的室内无柱空间和室外群组空间，对应于功能、行为、地形、观景及造境的需要。

　　在内外空间中清水混凝土的使用，凸显了反曲面悬索和鱼腹梁结构中的柱、梁、板构件，屋面排水构件及其组合延伸也成为建筑外观的组成部分，建筑结构构件之间镶砌了当地采集现场加工的毛石墙体，与基座的毛石墙体一起，是对当地渔村、石屋在材料、工法及地形处理上的学习和呼应。在混凝土和石材之间，玻璃和木板幕墙的使用增加了通透感，并柔化了建筑的冷峻表情。

摄影 / Photographer:
李季、李兴钢 / Li Ji, Li Xinggang

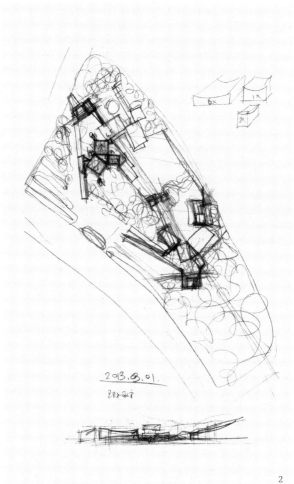

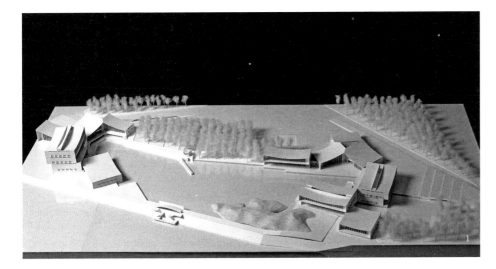

Initially designed in January 2013 and projected for completion and use in 2020, the Yuhuan Museum & Library is a complex that includes a library, a museum, and the public space in between the two buildings. It is located on reclaimed land and built in the new developmental zone of Yuhuan, an island county in Taizhou City, Zhejiang Province. The area of Kanmen in the old city of Yuhuan is unique, characterized by local fishing port. Kanmen includes diverse architectural and geographical elements such as a tunnel, a mountain, a port, a bay, a channel, and an island. The large landscape between the museum and library is a large body of water. Simultaneously, the overall spatial composition including the multi-level structures and the landscape is like "mountain-sea" scenery.

Both the museum and library were modeled as modern "fishing village settlements" with horizontal and vertical combination by curved roof building units. Both groups of buildings sit on large stone foundations accessible by long ramps and stairs. From the rear via a tunnel, visitors can enter the halls through the main entrance, alternatively enter the atrium at the back. These routes take visitors "through mountains as they enter the port". The units of the building are interlocked and connected in linearly or stacked vertically. The inner courtyards of the two buildings are relatively far away, and in opposite directions, connected by horizontal lawn of "water". Together with landscaping such as the fishing pavilion, wharf, dike, steps, and trees, they form an image of "fishing port" in the new developing area of Yuhuan. Borrowing the distant mountains and water, they are integrated to form "poetic scenery".

The project further develops a study of how to combine and integrate single, modular units to create buildings and group spaces. The museum and library were built using a concrete, cable suspension method to create a concave surface, and a similar large-span fish-belly beam structure were also used. The modular structures and units are repeatedly combined, varied, connected and enclosed in horizontal and vertical directions to form unique indoor column-free spaces and outdoor public spaces, corresponding to the needs of function, behavior of people, terrain, observation and situation.

The use of pale concrete in internal and external spaces highlights and demonstrates the significance of columns, beams and plates in the concave suspension and fish-belly beam structures. The roof drainage components and their combined extensions also become part of the buildings' appearance. The components in between building structures are inlaid with the local rubble walls processed on site. Together with the rubble walls of the base, they indicate an evolving attitude and response to construction methods and terrain treatment of the local fishing village and stone houses. Aside from concrete and stone, the use of glass and wood cladded walls creates a transparent appearance and softens the cold and hard expression of the buildings.

4 图书馆模型（1/200）/ Model of the library (1/200)
5 博物馆模型（1/200）/ Model of the museum (1/200)
6 结构空间单元模型（1/100）/ Model of the structure-space unit (1/100)
7 剖面模型（1/50）/ Section model (1/50)

4 图书馆模型（1/200）/ Model of the library (1/200)
5 博物馆模型（1/200）/ Model of the museum (1/200)
6 结构空间单元模型（1/100）/ Model of the structure-space unit (1/100)

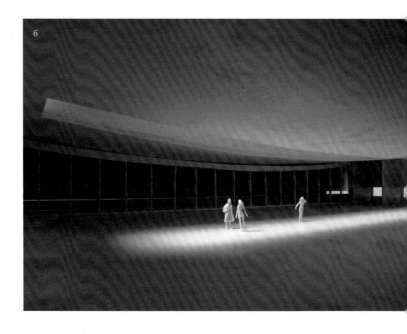

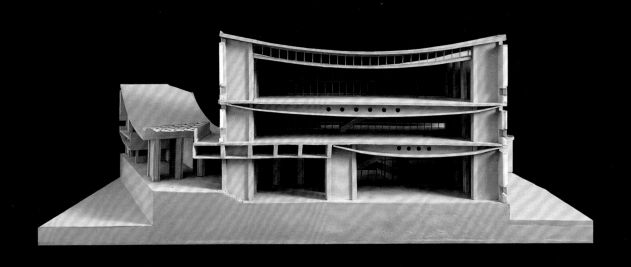

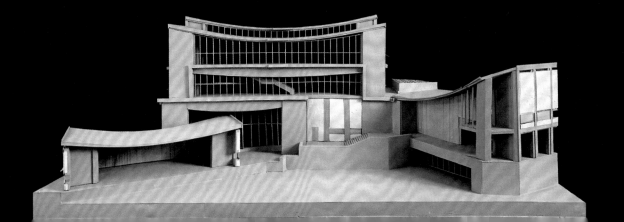

8

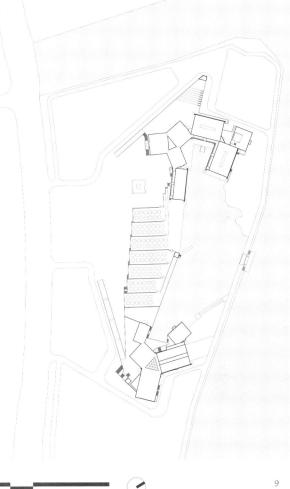

0 20 50   100m

9

0 2 5 10m

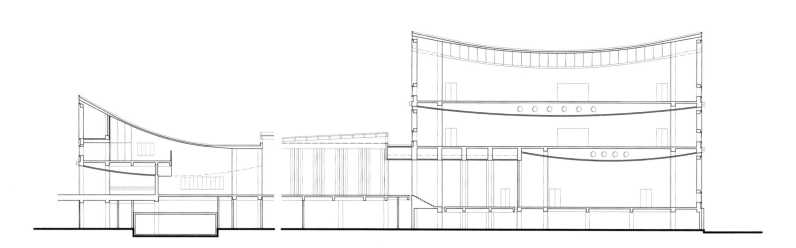

0   5   10   20m

1 博物馆门厅 / Entrance hall of the museum

2 综合大厅 / Comprehensive hall

3 临时展厅 / Temporarily exhibition hall

4 亭 / Pavilion

5 序言厅 / Preface hall

6 展厅 / Exhibition hall

7 多功能厅 / Multi-functional hall

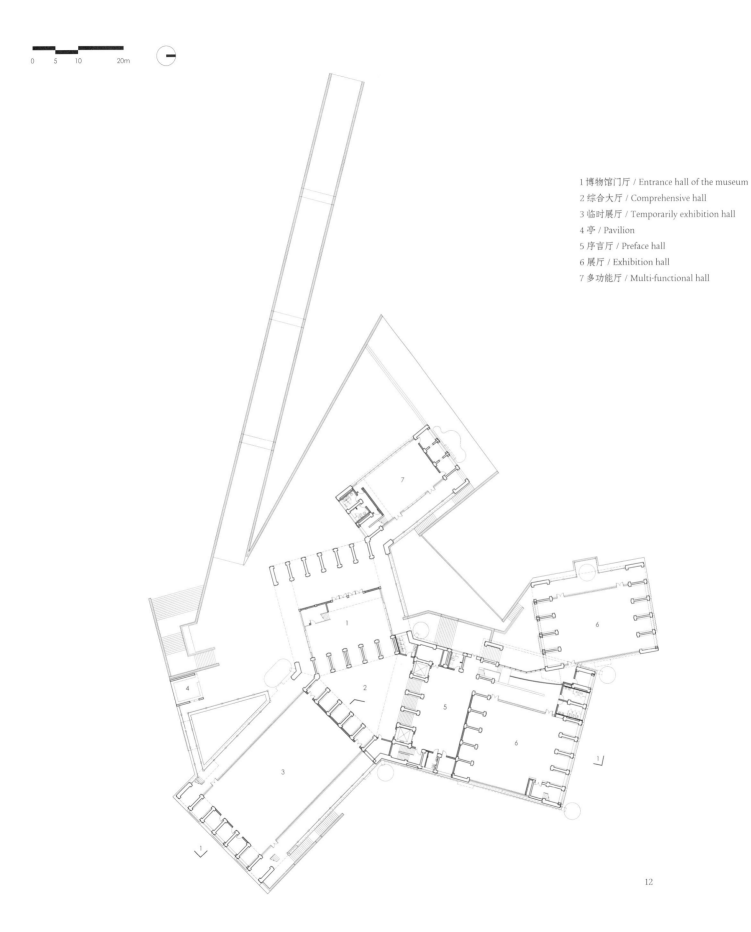

12

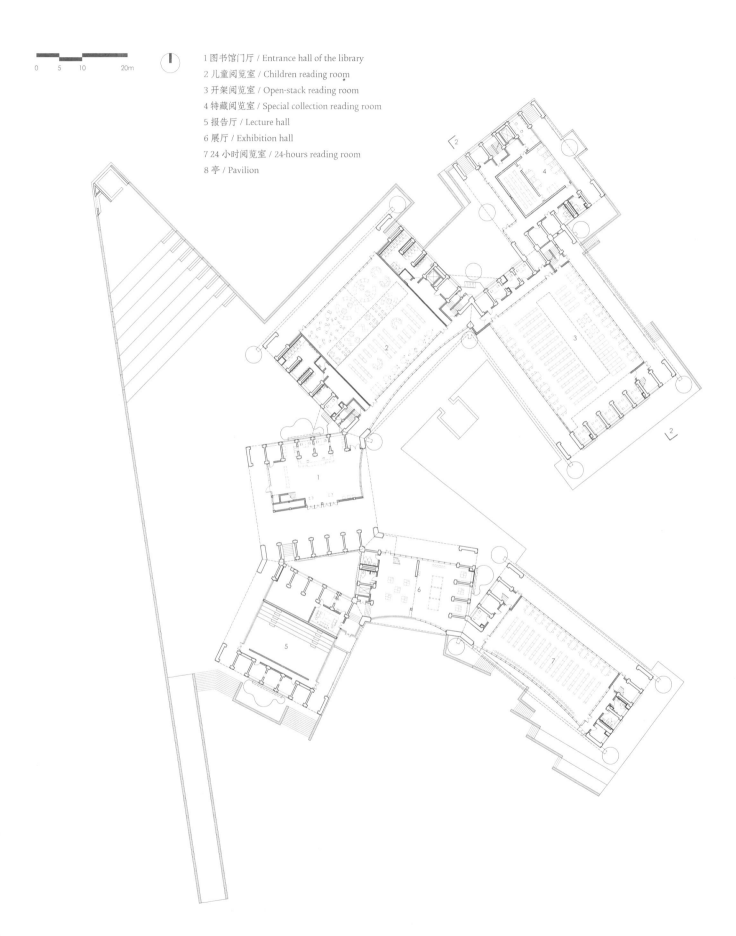

0   5   10   20m

1 图书馆门厅 / Entrance hall of the library
2 儿童阅览室 / Children reading room
3 开架阅览室 / Open-stack reading room
4 特藏阅览室 / Special collection reading room
5 报告厅 / Lecture hall
6 展厅 / Exhibition hall
7 24 小时阅览室 / 24-hours reading room
8 亭 / Pavilion

15

0 2   5   10m

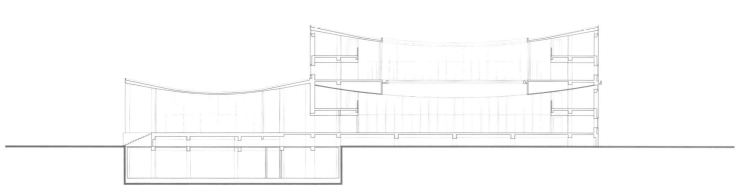

18

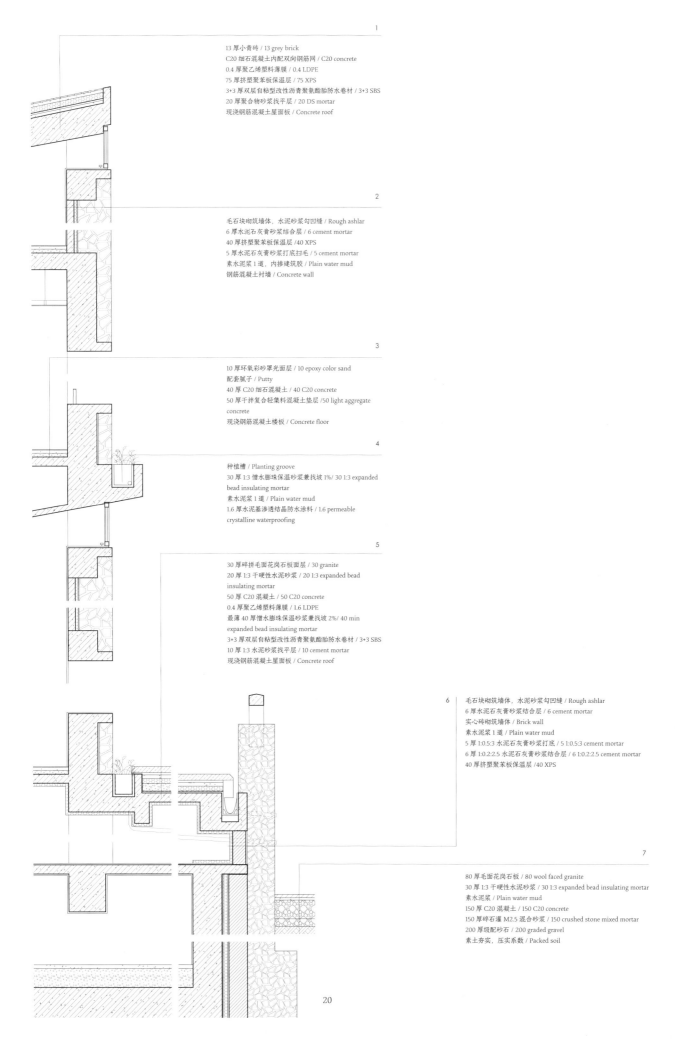

1

13 厚小青砖 / 13 grey brick
C20 细石混凝土内配双向钢筋网 / C20 concrete
0.4 厚聚乙烯塑料薄膜 / 0.4 LDPE
75 厚挤塑聚苯板保温层 / 75 XPS
3+3 厚双层自粘型改性沥青聚氨酯胎防水卷材 / 3+3 SBS
20 厚聚合物砂浆找平层 / 20 DS mortar
现浇钢筋混凝土屋面板 / Concrete roof

2

毛石块砌筑墙体，水泥砂浆勾凹缝 / Rough ashlar
6 厚水泥石灰膏砂浆结合层 / 6 cement mortar
40 厚挤塑聚苯板保温层 /40 XPS
5 厚水泥石灰膏砂浆打底扫毛 / 5 cement mortar
素水泥浆 1 道，内掺建筑胶 / Plain water mud
钢筋混凝土衬墙 / Concrete wall

3

10 厚环氧彩砂罩光面层 / 10 epoxy color sand
配套腻子 / Putty
40 厚 C20 细石混凝土 / 40 C20 concrete
50 厚干拌复合轻集料混凝土垫层 /50 light aggregate
concrete
现浇钢筋混凝土楼板 / Concrete floor

4

种植槽 / Planting groove
30 厚 1:3 憎水膨珠保温砂浆兼找坡 1%/ 30 1:3 expanded
bead insulating mortar
素水泥浆 1 道 / Plain water mud
1.6 厚水泥基渗透结晶防水涂料 / 1.6 permeable
crystalline waterproofing

5

30 厚碎拼毛面花岗石板面层 / 30 granite
20 厚 1:3 干硬性水泥砂浆 / 20 1:3 expanded bead
insulating mortar
50 厚 C20 混凝土 / 50 C20 concrete
0.4 厚聚乙烯塑料薄膜 / 1.6 LDPE
最薄 40 厚憎水膨珠保温砂浆兼找坡 2%/ 40 min
expanded bead insulating mortar
3+3 厚双层自粘型改性沥青聚氨酯胎防水卷材 / 3+3 SBS
10 厚 1:3 水泥砂浆找平层 / 10 cement mortar
现浇钢筋混凝土屋面板 / Concrete roof

6

毛石块砌筑墙体，水泥砂浆勾凹缝 / Rough ashlar
6 厚水泥石灰膏砂浆结合层 / 6 cement mortar
实心砖砌筑墙体 / Brick wall
素水泥浆 1 道 / Plain water mud
5 厚 1:0.5:3 水泥石灰膏砂浆打底 / 5 1:0.5:3 cement mortar
6 厚 1:0.2:2.5 水泥石灰膏砂浆结合层 / 6 1:0.2:2.5 cement mortar
40 厚挤塑聚苯板保温层 /40 XPS

7

80 厚毛面花岗石板 / 80 wool faced granite
30 厚 1:3 干硬性水泥砂浆 / 30 1:3 expanded bead insulating mortar
素水泥浆 / Plain water mud
150 厚 C20 混凝土 / 150 C20 concrete
150 厚碎石灌 M2.5 混合砂浆 / 150 crushed stone mixed mortar
200 厚级配砂石 / 200 graded gravel
素土夯实，压实系数 / Packed soil

20

2020

崇礼太子城雪花小镇
SNOWFLAKE TOWN
OF PRINCE CITY IN CHONGLI

崇礼太子城雪花小镇（北京冬奥会张家口赛区太子城冰雪小镇文创商街）设计始于 2017 年 11 月，将于 2020 年 10 月建成投入使用，位于北京 2022 冬奥会张家口赛区核心区的崇礼太子城冰雪小镇的中心地带，北倚太子城遗址，西临太子城高铁站，四面被群山环抱，遗址和高铁站的两条轴线交会于此，基地西北低、东南高，高差约 12 米。设计受精美绝伦的自然造物——冰晶雪花的启发，将其几何转换成抽象的雪花母题，在水平方向上错缝拼合，四向延伸，立体演变为雪花建筑及街市。天空中飘舞的"雪花"缓落于山谷，成簇成团，渐生出一个冬日山林中的冰雪世界。

不同于自上而下的城市规划体系，雪花小镇的生成法则是自下而上，通过几何原型的自然生长来完成的，是一个"自然聚落"。将雪花状的控制网格，以太子城和高铁站的轴线交点为中心，向四周扩散至整个场地。以几何切削的方式，来形成遗址轴线大道、颁奖广场等仪式空间；以摘掉"花瓣"的方式，在地块中获得街巷、院落等有机空间。顺应地形起伏，结合控制网格，将地下室底板和顶板设计为坡度 1.5% 西北低、东南高的整体斜板，"雪花建筑"落于斜板之上。每六个"花瓣"向心组合成一朵"雪花"，覆盖在"花心"高外缘低的坡屋顶下，通过脊线和檐口的设计，"雪花"可以连续拼合，"雪花"之间形成高差为 0.75 米的台地组团，每个组团由若干完整的"雪花"或不完整的"花瓣"拼合而成，组团之间形成丰富多样的内街和院落。区别于人工聚落明晰的组织层级，雪花小镇的室内和室外，具有模糊的空间界定和匀质的空间尺度。无论从山上俯瞰还是由地面人视，冰雪小镇舒展平缓的坡屋顶建筑匍匐于连续延伸的台地之上，因其标准单元的绵延组合而天然具有统一的建筑尺度和形象风貌，体现北方农耕文化和草原游牧文化的交融。

受益于六边体"雪花建筑"单元的重复组合和标准模数控制，建筑（地上）的结构体系和围护体系均可在工厂预制、现场装配完成。在户型、屋顶、立面等部位均按不同的标准化装配式单元设计，通过少量的模块类型进行大量、多样组合的方式，来实现建筑群体丰富多变的效果。

摄影 / Photographers:
中赫集团（提供）、李兴钢、王汉、易灵洁 / Sinobo Group, Li Xinggang, Wang Han, Yi Lingjie

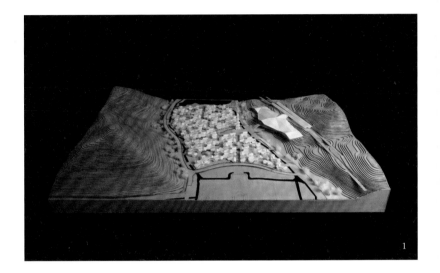

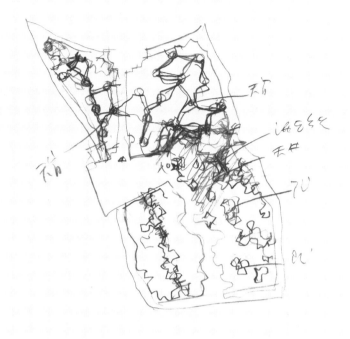

The Snowflake Town of Prince City in Chongli was initially designed in October 2017 and will be completed and put into use in 2020. It is located in the center of Chongli Prince City's Ice and Snow Town, the core area of Zhangjiakou competition zone for the Beijing 2022 Winter Olympic Games. Surrounded by mountains, it is close to the ruins of Prince City in the north and its high-speed railway station in the west. The two axes of the ruins and the high-speed railway station meet here. The site is low in the northwest and high in the southeast, with a height difference of about 12 meters. Inspired by the shape of a snowflake—a beautiful natural creation—the design transforms the site into an abstract snowflake motif. The Snowflake Town is staggered horizontally, extending in four directions that evolve into "snowflake buildings" and market streets in three dimensions. "Snowflakes" fluttering in the sky slowly fall in the valley and form clusters, gradually creating a world of ice and snow in the winter forest.

Different from the usual "top-down" planning process, we formed the Snowflake Town in a "bottom-up" way, through the natural growth of geometric prototype. It can be considered a "natural settlement". The snowflake form is used as controlling grid, centered at the intersection at the axes of Prince City and the high-speed railway station. Like its "natural" counterpart, the basic snowflake unit spreads around the site. The ceremonial spaces, including the axis avenue and the awards plaza, appear at the site and are formed through the geometric "cutting" of the snowflakes. The organic spaces such as the block streets and courtyards are obtained by removing "petals". Following the undulating terrain, we used controlling grid to create inclined basement floor and roof with a gradient of 1.5% (lower in the northwest and higher in southeast), upon which the "snowflake buildings" settle. The six "petals" of the snowflake merge at the center. Each "petal" is covered by a sloped roof radiating from high to low from the "snowflake's" center. Due to the design of the rooflines and cornices, the "snowflakes" can be joined to form terraced groups with a 0.75 meter height differential. Each group is composed of several complete "snowflakes" or incomplete "petals", forming a rich variety of inner streets and courtyards. Different from the usual hierarchy of planned settlements, the indoor and outdoor spaces of snowflake town has a less distinctive boundary and varied scale. Whether looking from the mountain or from the ground, it appears that the Snowflake Town stretches its gentle sloping roofs onto the terrace by continuous extension. Because the Snowflake Town's construction is based on the modular unit, it naturally has a unified architectural scale, identity, and style, which represents the integration of northern farming culture and grassland culture.

Due to the repeated combination of the standard, hexagonal snowflake shaped units, the building's structural and shelter system (above ground) can be prefabricated in factory and assembled on site. The plan, roof, and facade design is based on standardized assembly units. Despite our reliance on a limited number of snowflake modules, when used in combination with each other, we created a rich, diverse group of buildings.

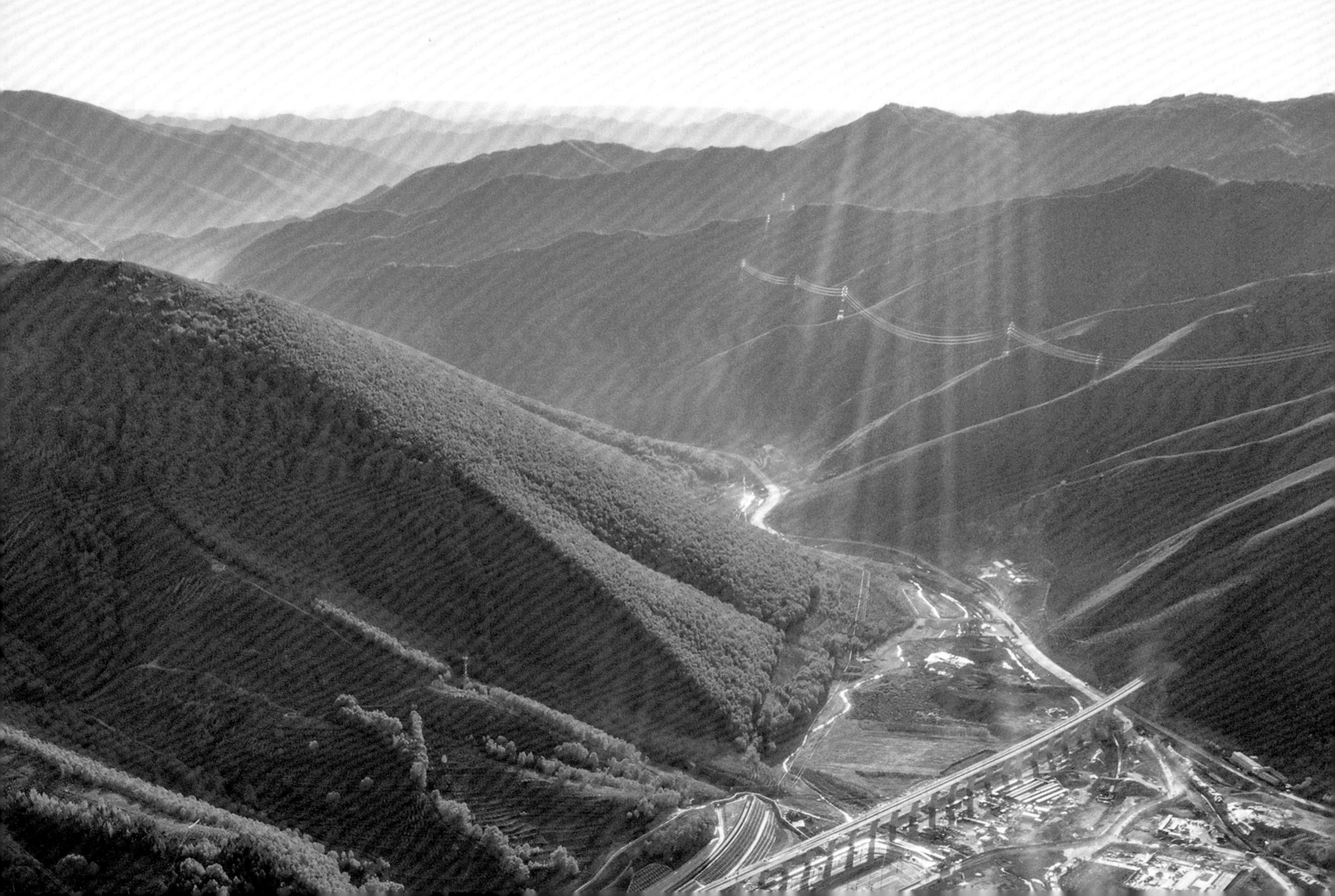

3 东南鸟瞰 / Aerial view from the southeast

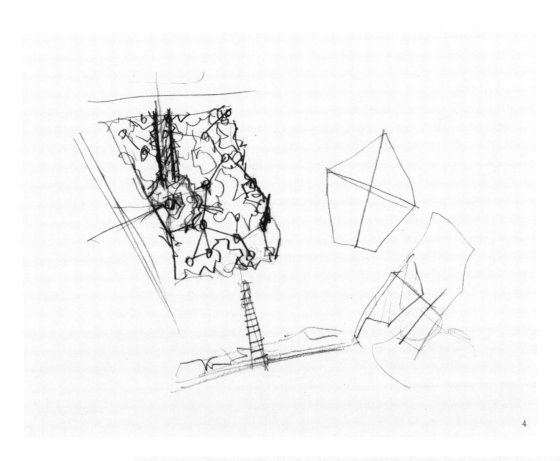

4

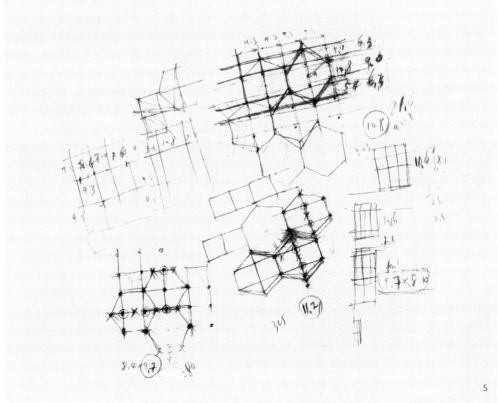

5

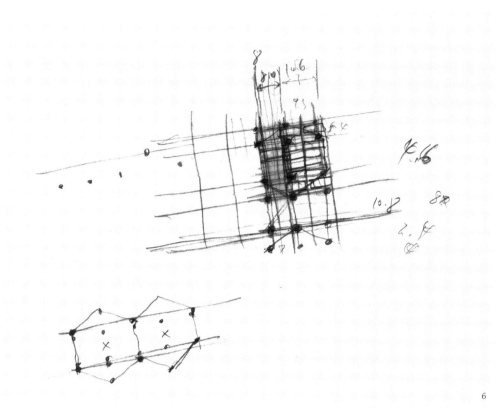

6

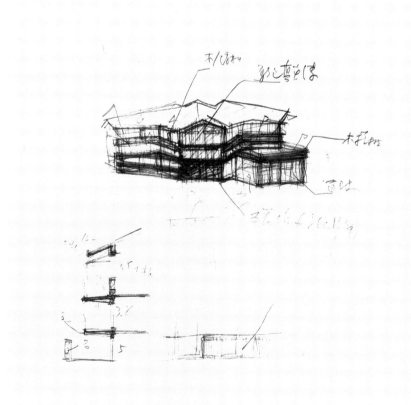

7

8-9 模型（1/500）/ Model (1/500)
10 结构单元模型（1/50）/ Model of the structure unit (1/50)
11 单元模型（1/50）/ Unit model (1/50)

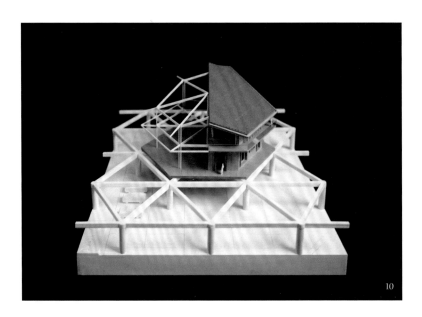

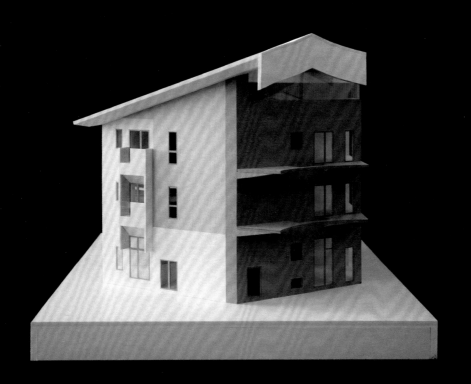

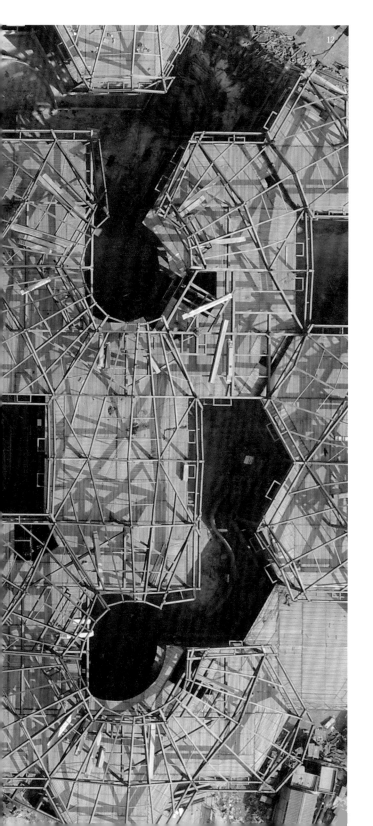

12

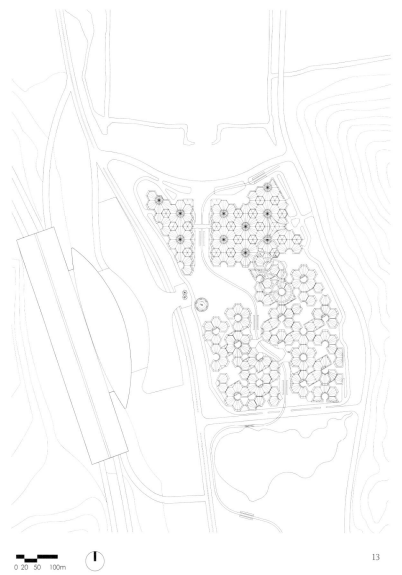

0 20 50 100m

13

481

1 奥运塔（未实施）/
Olympic tower (unimplemented)
2 颁奖广场 /Awards plaza
3 游览小火车 /Excursion train
4 轴线大道 / Axis avenue
5 商场 / Shopping mall
6 商业区 /Commercial area
7 民宿区 / B&B area

0 10 20   50m

14

0 5 10   30m

15

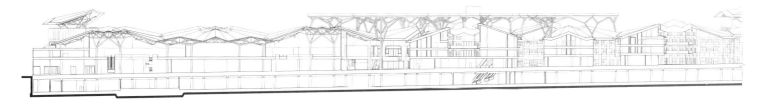

16

17

2020

北京 2022 年冬奥会与冬残奥会
延庆赛区场馆及设施
BEIJING 2022 OLYMPIC
& PARALYMPIC WINTER
GAMES YANQING ZONE
VENUES AND FACILITIES

北京 2022 年冬奥会与冬残奥会延庆赛区核心区规划和场馆及配套设施设计始于 2016 年 2 月，将于 2020 年竣工投入使用，位于燕山山脉军都山以南的小海坨南麓，风景秀丽，山高林密，地形复杂，用地狭促，是在设计上最具挑战性的北京冬奥赛区。延庆赛区的设计理念是 "山林场馆，生态冬奥"，通过建筑、景观和赛道设计的联合创新，打造冬奥场馆历史上新的里程碑，同时最大限度减少对环境的扰动，使建筑与自然景观有机结合，在满足呈现精彩赛事要求的基础上，建设一个融于自然山林中的冬奥赛区。

赛区南区的延庆冬奥村和国家雪车雪橇中心，隔着中部的山谷、河道和塘坝东西互望。国家雪车雪橇中心凌驾于山谷西侧的山脊之上和山林之间，若隐若现；延庆冬奥村则犹似一个山地村落，铺展在山谷东侧的台地之上，以分散式、半开放的院落格局，顺山形地势层层叠落，掩映于山林之中，中间有一处小庄科村遗迹被精心保留修缮，成为冬奥村独特的公共空间。

赛区北区是国家高山滑雪中心，竞速、竞技赛道及训练雪道由小海坨山顶向山谷蜿蜒而下，依托小海坨山的地形优势，以天然 "山石" 作为赛道主题要素，飞速滑行的 "动景" 给运动员和观众带来难忘的比赛、观赛体验；小海坨山最高点的山顶出发区，犹如凌空于山顶的巨大风筝；中间平台既是一个缆车换乘中心，又为观众提供绝佳的比赛观赏点；高山集散、媒体转播、各结束区等主要功能区，以珠链式布局散落在狭长险峻的山谷中，由预制装配式结构架设成为不同高度的错落平台，穿插叠落于山谷，减少对自然山体的改造，营造出与山地环境相得益彰的人工景观。除主要场馆之外，赛区内的众多市政公用设施——缆车站、变电站、输水泵站、造雪泵站、气象雷达站、管廊出入口等，也分别进行慎重选址和精心营造，使其适宜于所在的不同山地环境，并增强其公共性和观景功能。

充分利用区内自然地形环境和风景资源，结合各处特色场馆，构建立体山地景观格局，形成海坨飞鸢、晴雪揽胜、丹壁幽谷、凌水穿山、秋岭游龙、双村夕照、层台环翠、迎宾画廊等 "冬奥八景"，由一条贯穿整个园区的景观游览带串联起来，犹如在海坨山深谷实地营造一幅当代的大型山水图卷。

摄影 / Photographers:
张广源、孟阳、李兴钢、刘振 / Zhang Guangyuan, Meng Yang, Li Xinggang, Liu Zhen

486

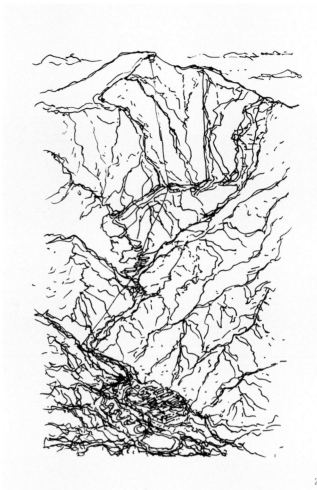

The Beijing 2022 Olympic & Paralympic Winter Games Yanqing Zone Venues and Facilities were initially planned and designed in January 2017 and will be completed and put into use in 2020. They are located at the southern foot of Xiaohaituo, south of Jundu Mountain in the Yanshan Mountains. Although situated on a site with beautiful scenery, tall mountains, and dense woods, this complicated terrain with very few flat lands to work with proved to be a most challenging Winter Olympic zone to design. The design concept of the Yanqing Zone is "Wooded Mountain Venues and Ecological Winter Olympics". With minimal disturbance to the environment, we created a new milestone in the history of Winter Olympic venue design. We used innovative architecture and organic design to integrate the natural landscape with the competition venues. Harmonizing the Olympic competition zone with the natural mountain forest would also result in venues conducive with the highest levels of elite competition.

The Yanqing Winter Olympic Village and National Sliding Center to the south of the competition zone face each other from east to west across the valley, over a dammed river and pond in between. The National Sliding Center is located above the mountain ridge on the west side of the valley. It is partly visible between the mountain forests. The Yanqing Winter Olympic Village is like a mountain village with a decentralized and semi-open courtyard pattern spread out on the terrace on the east side of the valley. It is situated along multi-layered mountain terrain and hidden in the mountain forest. In the middle of the village is the ruins of Xiaozhuangke Village, which has been carefully preserved and rehabilitated, becoming a special communal space for the Winter Olympic Village.

The north area of the competition zone is the National Alpine Ski Center. The downhill and slalom competition and training courses take advantage of Xiaohaituo Mountain's terrain, which meanders from the top of the mountain and into the valley. The natural "mountain stone" becomes the primary thematic element of the courses. The dynamic scene of rapid sliding brings unforgettable competition and watching experience to athletes and spectators. The starting area at the top of Xiaohaituo Mountain is similar to a huge kite flying over the top of the mountain. The middle platform is a cable car transfer station, providing spectators with an excellent view of the competition. The main transition areas for crowd gathering, exit, media and broadcast compound, and various finishing areas are scattered in a "string of pearls" layout in a narrow and deep valley. The prefabricated structural frames are set up as scattered platforms of different heights, which are interspersed and overlapped in the valley to reduce the physical changes to the mountain. Although artificial, the structures create a landscape that complements the mountain environment. In addition to the main venues, the infrastructure that supports the competition zone, such as cable stations, transformer substations, water delivery pump stations, snow making pump stations, weather radar stations, pipeline entry points and exits, are also carefully constructed to both make them suitable for different mountain environments based on careful site selections, and enhance their publicity and viewing functions.

In this project, we make full use of the existing, natural terrain and scenery in the zone, combining various characteristic venues to create a three-dimensional mountain landscape layout, and to represent the "Eight Winter Olympic Scenes": The Haituo flying kite, the scenic view in sunlit snow, the red cliff and deep valley, traversing water and crossing mountains, the wandering dragon on mountains in autumn, the twin mountain villages at sunset, the terraces with surrounding green trees, and the guest welcoming gallery. These scenes are connected by a landscape tour path throughout the park, forming a contemporary landscape painting in Haituo Mountains and its deep valleys.

3

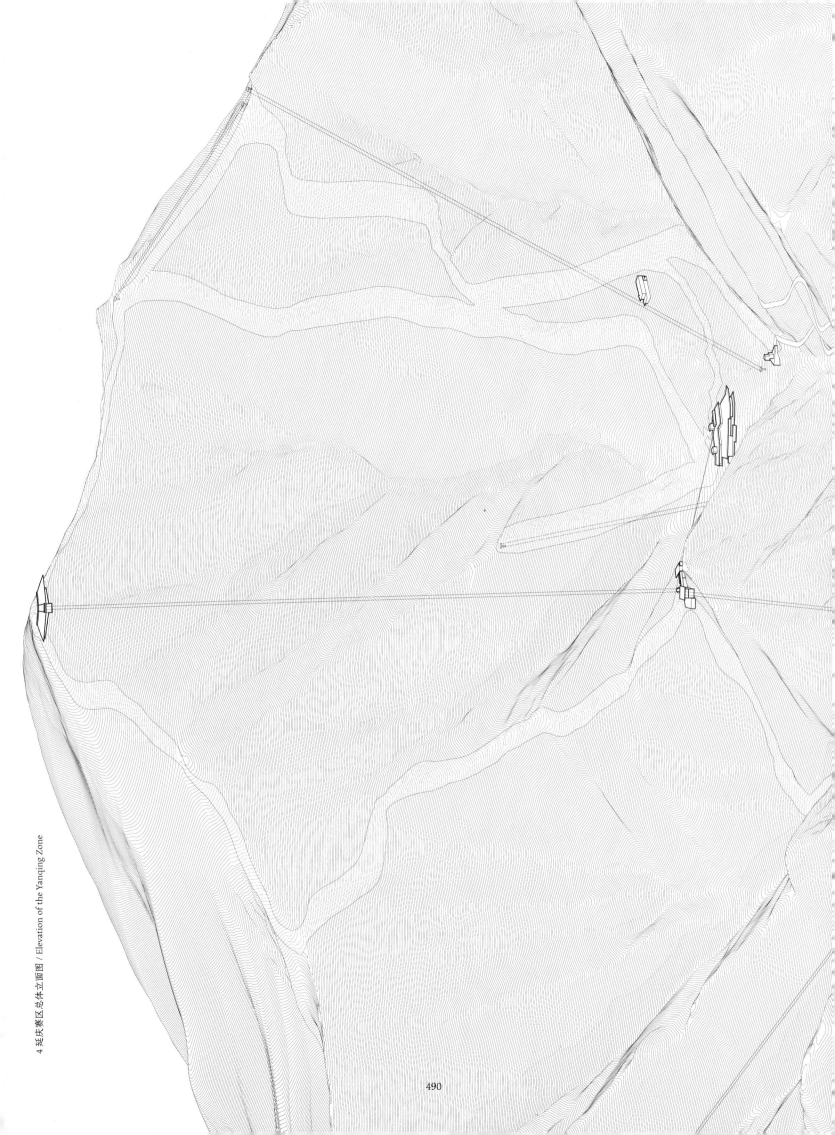

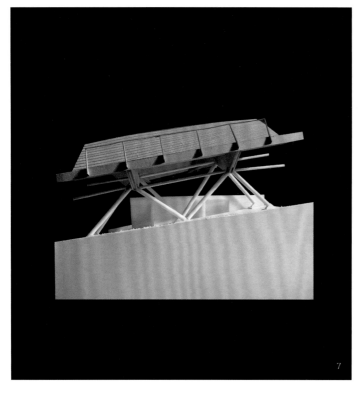

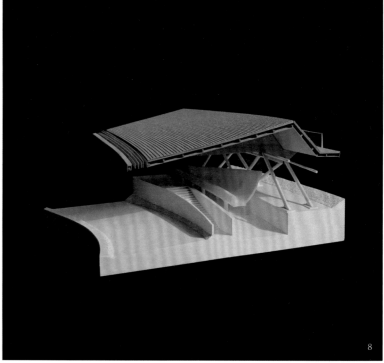

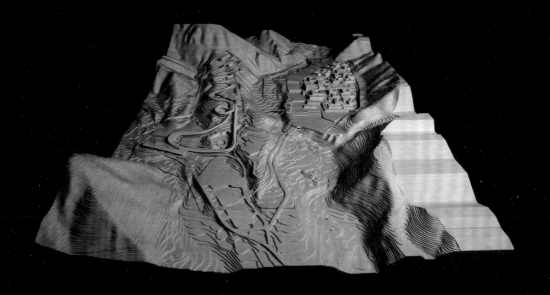

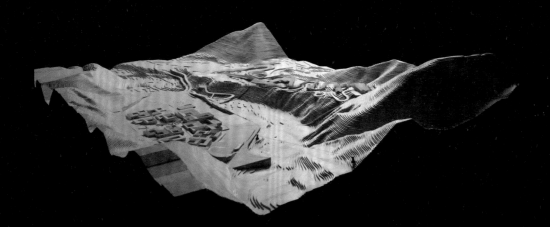

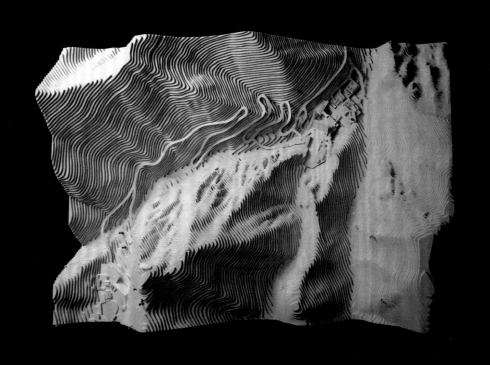
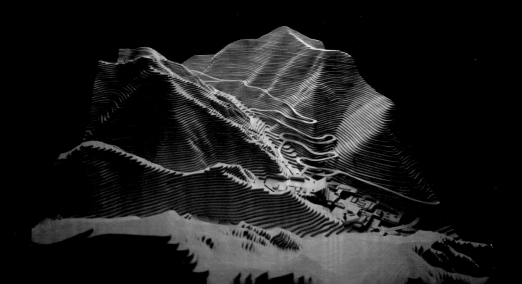

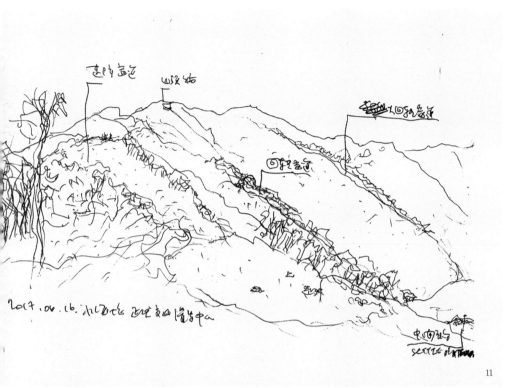

11

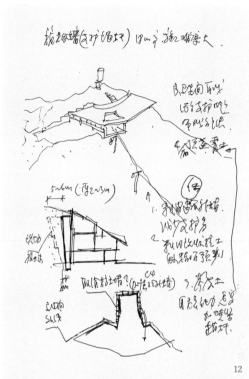

12

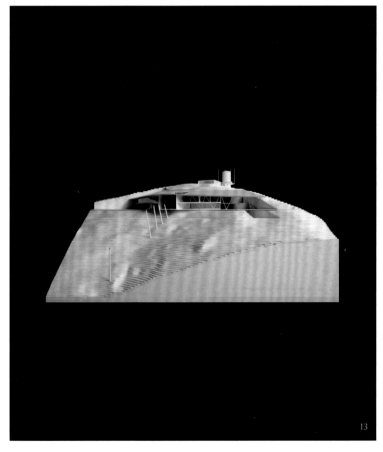

13

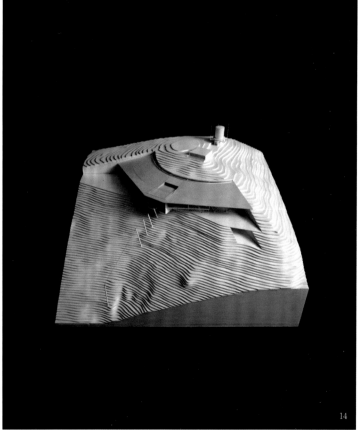

14

16

17

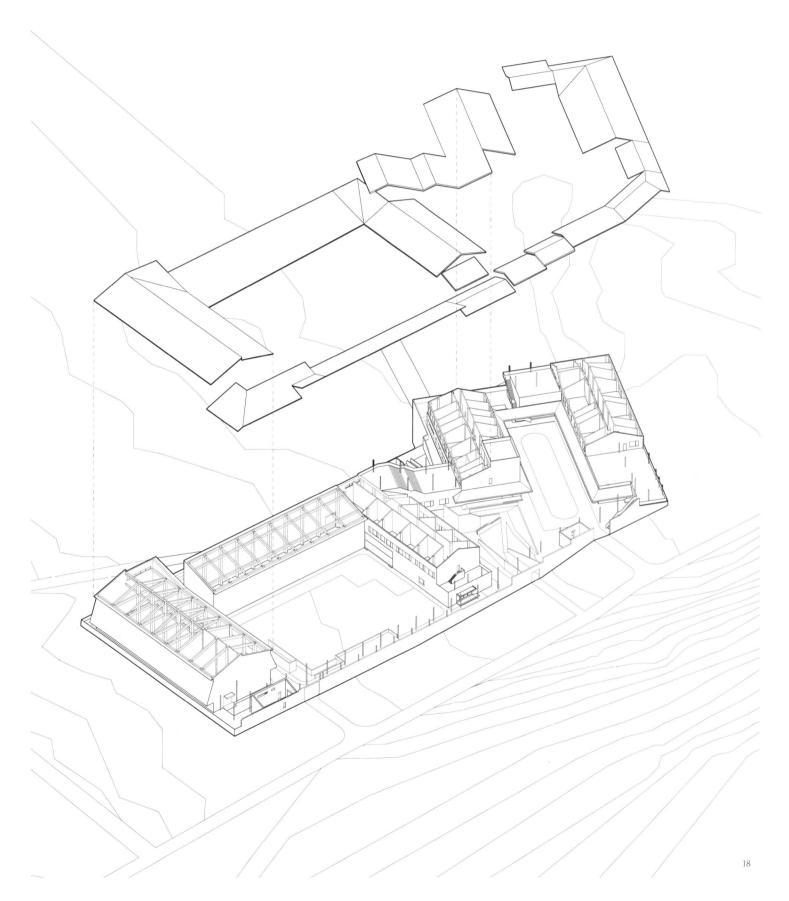

18

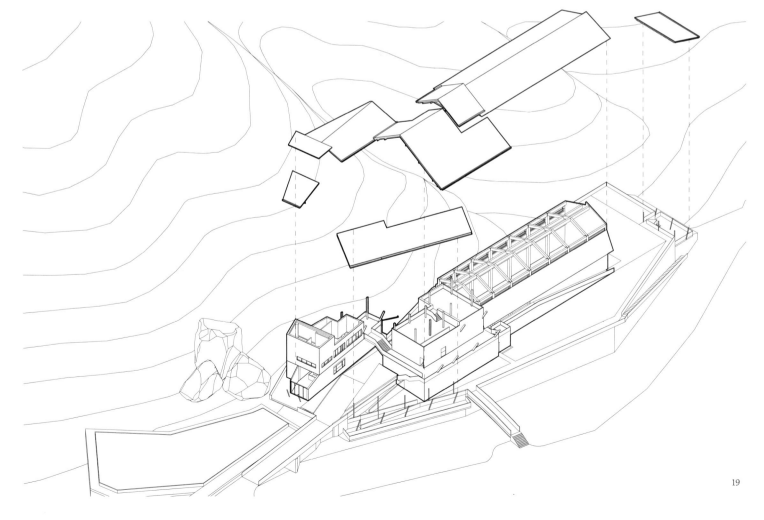

19

22 110KV 变电站东北侧人视 / View of the 110KV transformer substation from the northeast
23 110KV 变电站内部街厅人视 / View of the street-hall of the 110KV transformer substation

23

2015

瞬时桃花源
INSTANTANEOUS PEACH GARDEN

南京"瞬时桃花源"设计始于 2015 年 5 月，2015 年 7 月 8 日建成，同年 8 月 2 日拆除。作为南京大学—剑桥大学建筑与城市研究中心发起的"格物工作营"设计研究实施项目，"瞬时桃花源"是"胜景几何"理念的在地实验，是一种对"空间诗意"的探寻，一种对"房—山"这一山水画式的设计语言的尝试，以及一种基于身体尺度的"新模度"系统之可能性的思考。

变迁中的南京老城南一隅的花露岗区域，经由喧闹的集庆路，转至鸣羊街，依次经过胡家花园、古瓦官寺和一片未经使用的仿古建筑，周围城南民宅的残垣断壁越来越多，脚下的羊肠小路也开始变得杂草丛生。伴随着犬吠虫鸣，城市的喧嚣渐渐滤绝于耳，曲折尽致中，一个转身——人们来到一片"桃花源"。四组小建筑——"台阁""树亭""墙廊"和"山塔"结合场地中的台地、孤树、水池、废弃厂房、城墙以及大片的荒草麦田而建，与纷乱的城市现状对话，与场地尘封的历史对话。

基于脚手架快速施工及拆除的可能性，通过不同的连接和叠加方式，形成阁、亭、廊、塔等传统建筑"类型"的结构和空间骨架；以市井常见的黑色半透明遮阳网布形成必要的覆盖和围护。脚手架与人的尺度关联以及遮阳网布特有的半透明视觉特性，则赋予这组建筑与体验者之间密切生动而微妙的身体关联，并超越其材料物性和通常市井"美"的限制，营造出一种场地的、内在的诗意。通过一种"正在建造中"的临时感，映射当前的社会时代特征和城市的快速建造状态。标准脚手架的使用，是"乐高"系列装置的建筑化延续和拓展，是一种"新模度"的尝试：意即在不同规模、尺度的建筑及空间中，基于人的身体尺度而构成的模数化控制性工具和手段，并与建筑的空间、结构、形式、材料、建造特别是人的体验密切结合。

建筑施工耗时四天，留存不足一月，在拥有千年历史的花露岗场地里"上演"了一场短暂的"瞬时桃花源"，在南京古老城墙的见证下，这既是一个瞬时存在的桃花源，也是一种时代和生活的再现、再建和再见。

摄影 / Photographers:

孙海霆、李兴钢 / Sun Haiting, Li Xinggang

1 四个单体模型（1/200）/ Models of the four small buildings (1/200)
2 场地草图 / Site sketch
3 遮阳网布和脚手架 / Detail of the black sunshade cloth and the scaffold

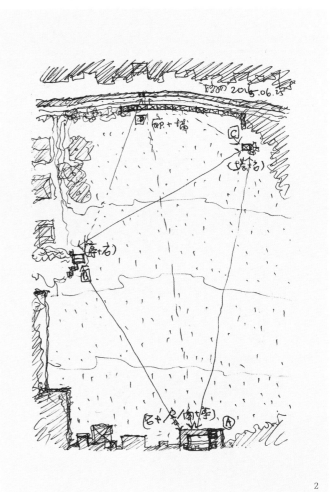

The Instantaneous Peach Garden in Nanjing was designed in May 2015, completed on July 8th 2015 and dismantled on August 2nd 2015. It is a design, research and implementation project for the Workshop of Gewu (Investigate-It) convened by University of Cambridge - Nanjing University Research Center on Architecture and Urbanism (CNRCAU). It can be considered a conceptual experiment of "Integrated Geometry and Poetic Scenery", an exploration of "spatial poetic", an interpretation of the landscape painting style design language of "room-mountain", and a thinking of the possibility of the "new modular" system based on bodily scale.

The route to the Peach Garden proceeds through the transforming Hualugang neighborhood in the southern corner of Nanjing's old city. Pedestrians turn towards Mingyang Street through the bustling Jiqing Road, and then pass the Hujia Garden, the Guwaguan Temple, and a block of unused buildings built in an ancient style. More and more houses in the south of the city are surrounded by ruins, and the narrow path is also becoming overgrown with weeds. Barking dogs and chirping insects filter out noise from the city. Walking through this road's twists and turns, people suddenly come upon the "Peach Garden". Four groups of small buildings—the Terrace Pavilion, the Tree Pavilion, the Wall Corridor, and the Mountain Tower—were built on the site, including a diverse set of elements, such as a ruin terrace, a solitary tree, a pool, abandoned factory buildings, the ancient city wall and the large field of grass and wheat, made dialogue with the city's present-day situation and site's history.

Relying on the possibility of rapid construction and demolition of the scaffold, the project created the structure and skeletal frame of traditional building types such as the terrace, pavilion, corridor and tower through different connection methods. The black sunshade cloth commonly seen in the city provided necessary cover and enclosure. The scale relation between scaffold and human body as well as the translucent visual characteristics of the sunshade net cloth enable visitors to experience close, vivid and subtle physical relations with buildings. We were able to transcend material limitations and common beauty concepts to create a sense of place and internal poetry. The project appears to be "under construction", reflecting the flux of current social conditions and rapid urban construction. The use of scaffolds is an extension of our studio's real-life architectural experiments of our study on the LEGO-type system to create modular structures. The new modularity considers space, structure, form, material, construction and people's experience on a human scale in different sized buildings and spaces.

The construction only took four days to complete and the project existed for less than one month. A transient "Instantaneous Peach Garden" was staged in the Hualugang site with over 1,000-year history. With the ancient city wall of Nanjing bearing witness, we created not only an instantaneous garden, but also developed a long lasting model of representation, of reconstruction, and a reflection of our life and times.

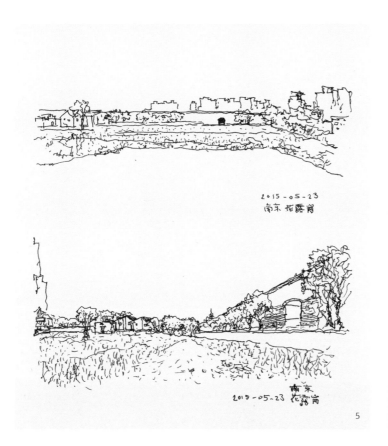

2015-05-23
南京 栖霞前

2015-05-23
南京 栖霞前

5

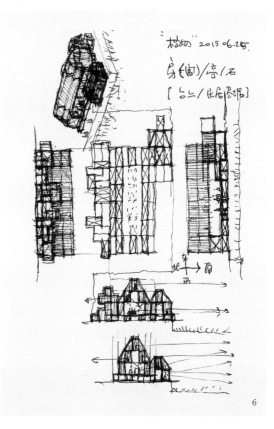

"栖霞" 2015.06.25.
另（侧）/亭/石
[与二/民居（院落）]

6

7

8

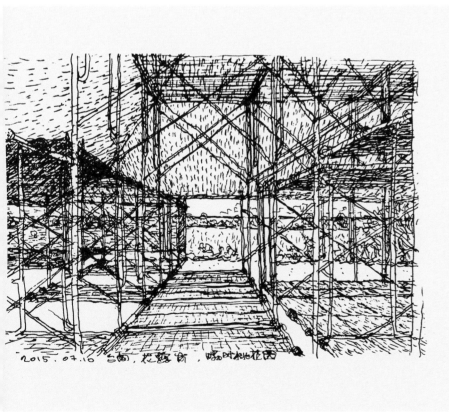

9

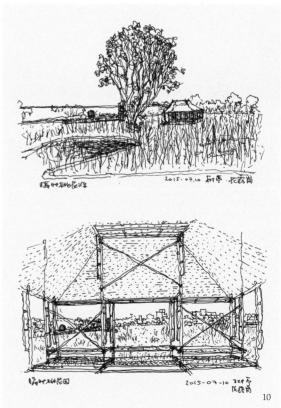

10

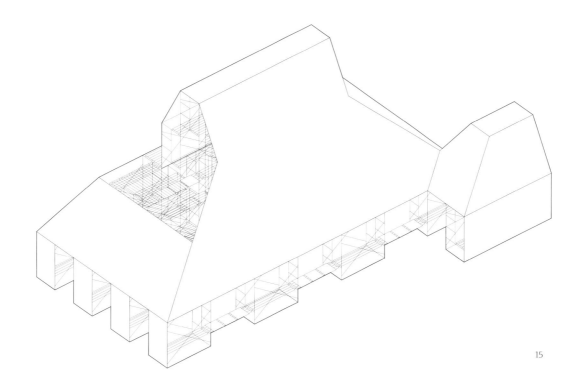

15

16

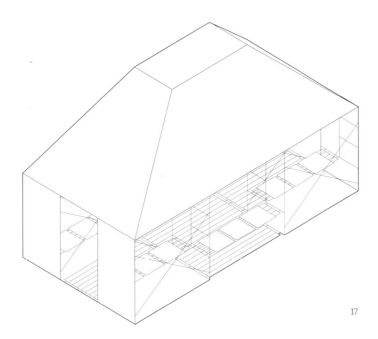

17

18

19

22

23

21

516

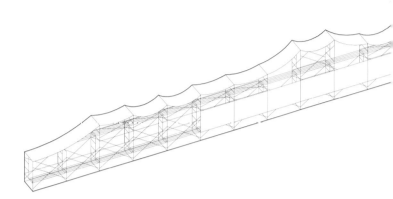

24

25

26

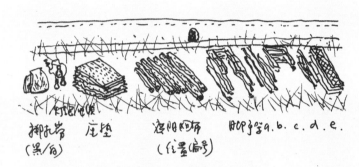

作品年表
CHRONOLOGY

兴涛社区（一期）●

Xingtao Residential Quarter (Step 1)

项目地点 / Location：北京 / Beijing

设计时间 / Design Period：1995.7—2000.4

施工时间 / Construction Period：1996.10—2002.10

用地面积 / Site Area：210,900 m²

建筑面积 / Floor Area：230,969 m²

建筑设计 / Architecture：李兴钢、高伟、林瞳、叶蕾、马先 / Li Xinggang, Gao Wei, Lin Tong, Ye Lei, Ma Xian

结构设计 / Structure：李淑捧、张晔、王立波 / Li Shupeng, Zhang Ye, Wang Libo

机电设计 / Engineering：夏树威、刘玉春、李颖 / Xia Shuwei, Liu Yuchun, Li Ying

北京兴涛学校 ●

Xingtao School in Beijing

项目地点 / Location：北京 / Beijing

设计时间 / Design Period：1996.5—1997.5

施工时间 / Construction Period：1997.6—1998.9

建筑面积 / Floor Area：27,000 m²

建筑设计 / Architecture：李兴钢、苗茁 / Li Xinggang, Miao Zhuo

结构设计 / Structure：李淑捧 / Li Shupeng

机电设计 / Engineering：夏树威、刘玉春、张福祥 / Xia Shuwei, Liu Yuchun, Zhang Fuxiang

新华大厦 ●

Xinhua Hotel

项目地点 / Location：河北唐山 / Tangshan, Hebei

设计时间 / Design Period：1998.3—1998.10

施工时间 / Construction Period：1998.10—2001.11

用地面积 / Site Area：5,400 m²

建筑面积 / Floor Area：18,852 m²

建筑设计 / Architecture：李兴钢、谭泽阳、于家峰、苏航、马先、张晖 / Li Xinggang, Tan Zeyang, Yu Jiafeng, Su Hang, Ma Xian, Zhang Hui

结构设计 / Structure：张晔 / Zhang Ye

机电设计 / Engineering：古晏、胡建丽、李炳华 / Gu Yan, Hu Jianli, Li Binghua

总图设计 / Master Plan：尤志毅 / You Zhiyi

室内设计 / Interior：张晖 / Zhang Hui

泰达小学 ●

Teda Primary School

项目地点 / Location：天津 / Tianjin

设计时间 / Design Period：1998.9—2000.6

施工时间 / Construction Period：2000.6—2001.9

用地面积 / Site Area：17,141 m²

建筑面积 / Floor Area：15,478 m²

建筑设计 / Architecture：李兴钢、胡水菁、张晔、宋淑辉、王宇 / Li Xinggang, Hu Shuijing, Zhang Ye, Song Shuhui, Wang Yu

结构设计 / Structure：刘连荣 / Liu Lianrong

机电设计 / Engineering：郑毅、胡建丽、马名东 / Zheng Yi, Hu Jianli, Ma Mingdong

室内设计 / Interior：张晔、盛燕 / Zhang Ye, Sheng Yan

兴涛会馆 ◣

Xingtao Club

项目地点 / Location：北京 / Beijing

设计时间 / Design Period：1999.7—2000.8

用地面积 / Site Area：9,214 m²

建筑面积 / Floor Area：10,629 m²

建筑设计 / Architecture：李兴钢、朱荷蒂、曾宁燕 / Li Xinggang, Zhu Hedi, Zeng Ningyan

北京西环广场暨西直门交通枢纽（与法国 AREP 合作）●

Xihuan Plaza & Xizhimen Transportation Exchange Hub (cooperated with AREP, France)

项目地点 / Location：北京 / Beijing

设计时间 / Design Period：2000.5—2007.4

施工时间 / Construction Period：2003.3—2008.3

用地面积 / Site Area：59,900 m²

建筑面积 / Floor Area：252,790 m²

建筑设计 / Architecture：崔愷、杜地阳、铁凯歌、李兴钢、苗茁、魏篙川、王宇、陈帅飞、于海为、刘爱华、张军英、金旭阳、宁霄、杜旭欧、申湘 / Cui Kai, Du Diyang, Tie Kaige, Li Xinggang, Miao Zhuo, Wei Gaochuan, Wang Yu, Chen Shuaifei, Yu Haiwei, Liu Aihua, Zhang Junying, Jin Xuyang, Ning Xiao, Du Xu'ou, Shen Xiang

结构设计 / Structure：吴平、彭永宏 / Wu Ping, Peng Yonghong

机电设计 / Engineering：夏树威、杨兰兰、刘海、宋孝春，刘继兴、王玉卿，贾京花、张福祥、陈琪、王烈 / Xia Shuwei，Yang Lanlan, Liu Hai, Song Xiaochun, Liu Jixing, Wang Yuqing, Jia Jinghua, Zhang Fuxiang, Chen Qi, Wang Lie

中关村生命科学园（竞赛方案）◣

Conference & Exhibition Center of Zhongguancun Biological Science Park（Competition）

项目地点 / Location：北京 / Beijing

设计时间 / Design Period：2001.3—2001.5

用地面积 / Site Area：62,379 m²

建筑面积 / Floor Area：58,653 m²

建筑设计 / Architecture：李兴钢、陈奕鹏、丁峰 / Li Xinggang, Chen Yipeng, Ding Feng

兴涛接待展示中心 ●

Xingtao Reception & Exhibition Center

项目地点 / Location：北京 / Beijing

设计时间 / Design Period：2001.4—2001.6

施工时间 / Construction Period：2001.6—2001.9

用地面积 / Site Area：2,510 m²

建筑面积 / Floor Area：883 m²

建筑设计 / Architecture：李兴钢、李靖、谭泽阳 / Li Xinggang, Li Jing, Tan Zeyang

结构设计 / Structure：王立波 / Wang Libo

机电设计 / Engineering：夏树威、刘玉春、张福祥 / Xia Shuwei, Liu Yuchun, Zhang Fuxiang

北京大兴区文化中心 ●
Daxing Culture Center
项目地点 / Location：北京 / Beijing
设计时间 / Design Period：2001.7—2004.9
施工时间 / Construction Period：2003.6—2006.6
用地面积 / Site Area：17,390 m²
建筑面积 / Floor Area：28,620 m²
建筑设计 / Architecture：李兴钢、陈晓唐、张晔、陈泽勇 / Li Xinggang, Chen Xiaotang, Zhang Ye, Chen Zeyong
结构设计 / Structure：孔雅莎 / Kong Yasha
机电设计 / Engineering：蔡力扬、冯小军、王晋恒 / Cai Liyang, Feng Xiaojun, Wang Jinheng
总图设计 / Master Plan：张黎 / Zhang Li
室内设计 / Interior：张晔、饶迈 / Zhang Ye, Rao Mai

东莞理工学院科研测试及后勤服务用房（一期）▲
Science Research Center & Service Facilities of Dongguan University of Technology (Step 1)
项目地点 / Location：广东东莞 / Dongguan, Guangdong
设计时间 / Design Period：2002.9—2003.1
施工时间 / Construction Period：2003.1—2005.11
用地面积 / Site Area：34,845 m²
建筑面积 / Floor Area：26,612 m²
建筑设计 / Architecture：李兴钢、付邦保、李大丹、肖晓丽、徐群宁（深圳华森建筑与工程设计顾问有限公司）/ Li Xinggang, Fu Bangbao, Li Dadan, Xiao Xiaoli, Xu Qunning (HSArchitects)
结构设计 / Structure：陈刘刚（深圳华森建筑与工程设计顾问有限公司）/ Chen Liugang (HSArchitects)
机电设计 / Engineering：王益、李瑞珍、杨虎（深圳华森建筑与工程设计顾问有限公司）/ Wang Yi, Li Ruizhen, Yang Hu (HSArchitects)

国家体育场——2008 年奥运会主体育场（与瑞士 Herzog & de Meuron 建筑事务所团队合作）●
The National Stadium — the Main Stadium of the 2008 Olympic Games (cooperated with Herzog & de Meuron Architects, Switzerland)
项目地点 / Location：北京 / Beijing
设计时间 / Design Period：2003.1—2005.6
施工时间 / Construction Period：2003.12—2008.4
用地面积 / Site Area：204,124 m²
建筑面积 / Floor Area：258,000 m²
建筑设计 / Architecture：Pierre de Meuron、Jacque Herzog、李兴钢、Herzog & de Meuron 建筑师团队、谭泽阳、邱涧冰、安鹏、张军英 / Pierre de Meuron, Jacque Herzog, Li Xinggang, Team of Herzong & de Meuron, Tan Zeyang, Qiu Jianbing, An Peng, Zhang Junying
结构设计 / Structure：Arup 结构顾问团队、范重、尤天直、胡纯炀、唐杰、王大庆 / Team of Arup, Fan Zhong, You Tianzhi, Hu Chunyang, Tang Jie, Wang Daqing
机电设计 / Engineering：丁高、郭汝艳、刘鹏、胡建丽、王玉卿、王健 / Ding Gao, Guo Ruyan, Liu Peng, Hu Jianli, Wang Yuqing, Wang Jian
室内设计 / Interior：谈星火 / Tan Xinghuo

东莞松山湖文化营多功能活动中心 ▲
Multi-Center of Songshanhu Culture Camp
项目地点 / Location：广东东莞 / Dongguan, Guangdong
设计时间 / Design Period：2003.8—2003.11
建筑面积 / Floor Area：17,130m²
建筑设计 / Architecture：李兴钢、付邦保、张音玄 / Li Xinggang, Fu Bangbao, Zhang Yinxuan

湖北省艺术馆（竞赛方案）▲
Hubei Art Gallery (Competition)
项目地点 / Location：湖北武汉 / Wuhan, Hubei
设计时间 / Design Period：2003.11—2004.1
用地面积 / Site Area：20,427 m²
建筑面积 / Floor Area：22,654 m²
建筑设计 / Architecture：李兴钢、付邦保、张音玄 / Li Xinggang, Fu Bangbao, Zhang Yinxuan

建川镜鉴博物馆暨汶川地震纪念馆 ●
Jianchuan Mirror Museum & Wenchuan Earthquake Memorial
项目地点 / Location： 四川大邑 / Dayi, Sichuan
设计时间 / Design Period： 2004.2—2009.12
施工时间 / Construction Period： 2004.8—2010.9
用地面积 / Site Area： 3,847 m²
建筑面积 / Floor Area： 6,098 m²
建筑设计 / Architecture： 李兴钢、谭泽阳、刘爱华、张音玄、付邦保 / Li Xinggang, Tan Zeyang, Liu Aihua, Zhang Yinxuan, Fu Bangbao
结构设计 / Structure： 王力波 / Wang Libo
机电设计 / Engineering： 夏树威、李超英、甄毅 / Xia Shuwei, Li Chaoying, Zhen Yi
总图设计 / Master Plan： 齐海娟 / Qi Haijuan

北京复兴路乙 59-1 号改造 ●
Renovation of No. B-59-1, Fuxing Road in Beijing
项目地点 / Location： 北京 / Beijing
设计时间 / Design Period： 2004.10—2005.8
施工时间 / Construction Period： 2005.7—2007.5
用地面积 / Site Area： 1,280 m²
建筑面积 / Floor Area： 5,402 m²
建筑设计 / Architecture： 李兴钢、张音玄、付邦保、谭泽阳 / Li Xinggang, Zhang Yinxuan, Fu Bangbao, Tan Zeyang
结构设计 / Structure： 蒋航军 / Jiang Hangjun
机电设计 / Engineering： 安岩、刘冰、宋晓梅 / An Yan, Liu Bing, Song Xiaomei

中华英才半月刊社综合业务楼 ●
Office Building of Top China Microdevices
项目地点 / Location： 北京 / Beijing
设计时间 / Design Period： 2005.12—2006.2
用地面积 / Site Area： 4,528 m²
建筑面积 / Floor Area： 13,332 m²
建筑设计 / Architecture： 李兴钢、张音玄、付邦保、郭佳 / Li Xinggang, Zhang Yinxuan, Fu Bangbao, Guo Jia

威海 "Hiland · 名座" ●
"Hiland · Mingzuo" in Weihai
项目地点 / Location： 山东威海 / Weihai, Shandong
设计时间 / Design Period： 2006.1—2007.9
施工时间 / Construction Period： 2007.10—2013.11
用地面积 / Site Area： 3,618 m²
建筑面积 / Floor Area： 26,190m²
建筑设计 / Architecture： 李兴钢、谭泽阳、钟鹏、肖育智、李宁 / Li Xinggang, Tan Zeyang, Zhong Peng, Xiao Yuzhi, Li Ning
结构设计 / Structure： 尤天直、唐杰、高文军、许庆 / You Tianzhi, Tang Jie, Gao Wenjun, Xu Qing
机电设计 / Engineering： 陶涛、王加、王玉卿、王莉 / Tao Tao, Wang Jia, Wang Yuqing, Wang Li

乳山文博中心 ●
Rushan Culture Center & Museum
项目地点 / Location： 山东乳山 / Rushan, Shandong
设计时间 / Design Period： 2006.7—2007.6
用地面积 / Site Area： 560,000 m²
建筑面积 / Floor Area： 24,000 m²
建筑设计 / Architecture： 李兴钢、张哲、钟鹏、弓蒙 / Li Xinggang, Zhang Zhe, Zhong Peng, Gong Meng

崀园 ◤
Keyuan Garden
项目地点 / Location：广西南宁 / Nanning, Guangxi
设计时间 / Design Period：2006.12—2008.1
用地面积 / Site Area：3,389 m²
建筑面积 / Floor Area：1,246 m²
建筑设计 / Architecture：李兴钢、付邦保、张哲、郭佳 / Li Xinggang, Fu Bangbao, Zhang Zhe, Guo Jia

上海世博会中国馆（竞赛方案）◤
China Pavilion in EXPO 2010 Shanghai (Competition)
项目地点 / Location：上海 / Shanghai
设计时间 / Design Period：2007.5—2007.8
用地面积 / Site Area：65,200 m²
建筑面积 / Floor Area：86,430 m²
建筑设计 / Architecture：李兴钢、张音玄、张哲 / Li Xinggang, Zhang Yinxuan, Zhang Zhe

乐高1号、乐高2号 ●
Lego I & Lego II
项目地点 / Location：北京 / Beijing
设计时间 / Design Period：2007.4、2008.3
施工时间 / Construction Period：2007.7、2008.4
作品尺寸 / Size：400mm × 500mm × 1200mm, 1800mm × 1000mm × 1200mm
建筑设计 / Architecture：李兴钢、张音玄、付邦保、张哲、郭佳、李宁、邢迪、张玉婷 / Li Xinggang, Zhang Yinxuan, Fu Bangbao, Zhang Zhe, Guo Jia, Li Ning, Xing Di, Zhang Yuting

唐山地震遗址纪念公园（竞赛方案）◤
Tangshan Earthquake Site Memorial Park (Competition)
项目地点 / Location：河北唐山 / Tangshan, Hebei
设计时间 / Design Period：2007.5—2007.7
用地面积 / Site Area：400,000 m²
建筑设计 / Architecture：李兴钢、郭佳、张哲、李宁、弓蒙 / Li Xinggang, Guo Jia, Zhang Zhe, Li Ning, Gong Meng

中国海关博物馆（与场域建筑工作室及九源三星建筑师事务所合作）●
China Custom Museum (cooperated with Approach Architecture Studio & Jiuyuan Samsung Architects)
项目地点 / Location：北京 / Beijing
设计时间 / Design Period：2007.6—2009.10
施工时间 / Construction Period：2009.11—2014.10
用地面积 / Site Area：21,023 m²
建筑面积 / Floor Area：33,000 m²
建筑设计 / Architecture：李兴钢、场域建筑工作室（梁井宇、彭小虎）、付邦保、肖育智、九源三星建筑师事务所（张安星、江曼、王松青）/ Li Xinggang, Approach Architecture Studio (Liang Jingyu, Peng Xiaohu), Fu Bangbao, Xiao Yuzhi, Jiuyuan Samsung Architecs (Zhang Anxing, Jiang Man, Wang Songqing)
结构设计 / Structure：李欣（九源三星建筑师事务所）/ Li Xin (Jiuyuan Samsung Architecs)
机电设计 / Engineering：雷昊、宋小亮、刘伊、王皓（九源三星建筑师事务所）/ Lei Hao, Song Xiaoliang, Liu Yi, Wang Hao (Jiuyuan Samsung Architects)

武汉市档案馆暨城建档案馆（竞赛方案）◤
Wuhan Archives (Competition)
项目地点 / Location：湖北武汉 / Wuhan, Hubei
设计时间 / Design Period：2007.8
用地面积 / Site Area：37,500 m²
建筑面积 / Floor Area：57,500 m²
建筑设计 / Architecture：李兴钢、付邦保、邢迪、张哲 / Li Xinggang, Fu Bangbao, Xing Di, Zhang Zhe

深圳湾体育中心（竞赛方案）
Shenzhen Bay Sports Center (Competition)
项目地点 / Location：广东深圳 / Shenzhen, Guangdong
设计时间 / Design Period：2007.9—2007.11
用地面积 / Site Area：307,740 m²
建筑面积 / Floor Area：268,621 m²
建筑设计 / Architecture：李兴钢、付邦保、钟鹏、李宁、邢迪、王子耕、谭泽阳 / Li Xinggang, Fu Bangbao, Zhong Peng, Li Ning, Xing Di, Wang Zigeng, Tan Zeyang

藕园
Couple Garden
项目地点 / Location：广东番禺 / Panyu, Guangdong
设计时间 / Design Period：2007.10—2008.1
用地面积 / Site Area：985 m²
建筑面积 / Floor Area：1,270 m²
建筑设计 / Architecture：李兴钢、付邦保、邢迪 / Li Xinggang, Fu Bangbao, Xing Di

威尼斯纸砖房 ●
Paper-brick House in Venice
项目地点 / Location：意大利威尼斯 / Venice, Italy
设计时间 / Design Period：2008.5
施工时间 / Construction Period：2008.9
作品尺寸 / Size：15m × 2m × 4m
建筑设计 / Architecture：李兴钢、付邦保、李宁、孙鹏 / Li Xinggang, Fu Bangbao, Li Ning, Sun Peng

李兴钢工作室室内 ●
Interior Design of Atelier Li Xinggang
项目地点 / Location：北京 / Beijing
设计时间 / Design Period：2008.7—2008.9
施工时间 / Construction Period：2008.9—2009.2
建筑面积 / Floor Area：485m²
建筑设计 / Architecture：李兴钢、郭佳、张玉婷、邱涧冰、谭泽阳 / Li Xinggang, Guo Jia, Zhang Yuting, Qiu Jianbing, Tan Zeyang

北京地铁 4 号线及大兴线地面出入口及附属设施 ●
Entrances of Line 4 & Daxing Line of the Beijing Subway
项目地点 / Location：北京 / Beijing
设计时间 / Design Period：2008.8—2009.3、2010.2—2010.4
施工时间 / Construction Period：2009.3—2009.10、2010.7—2010.12
建筑面积 / Floor Area：26,190 m²、8,371 m²
建筑设计 / Architecture：李兴钢、邱涧冰、肖宇智、李宁、邢迪、张玉婷、闫昱、唐勇 / Li Xinggang, Qiu Jianbing, Xiao Yuzhi, Li Ning, Xing Di, Zhang Yuting, Yan Yu, Tang Yong
结构设计 / Structure：高银鹰 / Gao Yinying
机电设计 / Engineering：许士骅 / Xu Shihua
总图设计 / Master Plan：余晓东 / Yu Xiaodong

北京地铁昌平线西二旗站 ●
Xi'erqi Station of Changping Line of the Beijing Subway
项目地点 / Location：北京 / Beijing
设计时间 / Design Period：2008.11—2010.4
施工时间 / Construction Period：2009.10—2010.12
建筑面积 / Floor Area：16,670m²
建筑设计 / Architecture：李兴钢、白芳、张音玄、李慧琴、赵国璆、张哲、张玉婷 / Li Xinggang, Bai Fang, Zhang Yinxuan, Li Huiqin, Zhao Guoqiu, Zhang Zhe, Zhang Yuting
结构设计 / Structure：王立波 / Wang Libo
机电设计 / Engineering：韦航、许士骅、杨兰兰 / Wei Hang, Xu Shihua, Yang Lanlan

商丘博物馆 ●
Shangqiu Museum
项目地点 / Location：河南商丘 / Shangqiu, Henan
设计时间 / Design Period：2008.12—2010.9
施工时间 / Construction Period：2011.3—2015.5
用地面积 / Site Area：73,613 m²
建筑面积 / Floor Area：29,672 m²
建筑设计 / Architecture：李兴钢、付邦保、谭泽阳、张哲、李喆、张玉婷、梁旭 / Li Xinggang, Fu Bangbao, Tan Zeyang, Zhang Zhe, Li Zhe, Zhang Yuting, Liang Xu
结构设计 / Structure：张晔 / Zhang Ye
机电设计 / Engineering：陈宁、刘燕军、王铮 / Chen Ning, Liu Yanjun, Wang Zheng
景观设计 / Landscape：李力 / Li Li

唐山"第三空间" ●
The "Third Space" in Tangshan
项目地点 / Location：河北唐山 / Tangshan, Hebei
设计时间 / Design Period：2009.2—2010.9
施工时间 / Construction Period：2010.6—2015.6
用地面积 / Site Area：12,852 m²
建筑面积 / Floor Area：88,011 m²
建筑设计 / Architecture：李兴钢、付邦保、孙鹏、赵小雨、谭泽阳、张一婷 / Li Xinggang, Fu Bangbao, Sun Peng, Zhao Xiaoyu, Tan Zeyang, Zhang Yiting
结构设计 / Structure：张付奎、孔文华 / Zhang Fukui, Kong Wenhua
机电设计 / Engineering：赵昕、李建业、胡建丽、王微微、许冬梅 / Zhao Xin, Li Jianye, Hu Jianli, Wang Weiwei, Xu Dongmei
照明顾问 / Lighting：郑见伟 / Zheng Jianwei

隆福寺项目（北京地铁六号线东四站织补工程） ◤
Longfusi Project (Commercial Facilities of Dongsi Station of Line 6 of Beijing Subway)
项目地点 / Location：北京 / Beijing
设计时间 / Design Period：2009.3—2019.12
施工时间 / Construction Period：2019.12—
用地面积 / Site Area：12,009 m²
建筑面积 / Floor Area：50,681 m²
建筑设计 / Architecture：邱涧冰、闫昱、侯新觉、张玉婷、张音玄、肖育智、张哲、李喆、钟曼琳 / Qiu Jianbing, Yan Yu, Hou Xinjue, Zhang Yuting, Zhang Yinxuan, Xiao Yuzhi, Zhang Zhe, Li Zhe, Zhong Manlin
结构设计 / Structure：曹清、郑红卫 / Cao Qing, Zheng Hongwei
机电设计 / Engineering：刘海、刘洞阳、汪春华、李金双、肖彦、蒲域 / Liu Hai, Liu Dongyang, Wang Chunhua, Li Jinshuang, Xiao Yan, Pu Yu
总图设计 / Master Plan：刘晓琳 / Liu Xiaolin

元上都遗址博物馆 ●
Museum for Site of Xanadu
项目地点 / Location：内蒙古正蓝旗 / Zhenglanqi, Inner Mongolia
设计时间 / Design Period：2009.5—2010.7
施工时间 / Construction Period：2010.7—2015.12
用地面积 / Site Area：6,747 m²
建筑面积 / Floor Area：4,997 m²
建筑设计 / Architecture：李兴钢、谭泽阳、赵小雨 / Li Xinggang, Tan Zeyang, Zhao Xiaoyu
结构设计 / Structure：王力波、高银鹰、张剑涛 / Wang Libo, Gao Yinying, Zhang Jiantao
机电设计 / Engineering：刘海、何猛、李超英、向波、甄毅 / Liu Hai, He Meng, Li Chaoying, Xiang Bo, Zhen Yi

海南国际会展中心 ●
Hainan International Convention & Exhibition Center
项目地点 / Location：海南海口 / Haikou, Hainan
设计时间 / Design Period：2009.7—2010.3
施工时间 / Construction Period：2009.11—2011.7
用地面积 / Site Area：319,873 m²
建筑面积 / Floor Area：132,788 m²
建筑设计 / Architecture：李兴钢、谭泽阳、付邦保、张玉婷 / Li Xinggang, Tan Zeyang, Fu Bangbao, Zhang Yuting
结构设计 / Structure：任庆英、王载、王文宇、谷昊 / Ren Qingying, Wang Zai, Wang Wenyu, Gu Hao
机电设计 / Engineering：郭汝艳、吴连荣、孙淑莉、郑坤、金健、曹磊、王浩然、张青、王健 / Guo Ruyan, Wu Lianrong, Sun Shuli, Zheng Kun, Jin Jian, Cao Lei, Wang Haoran, Zhang Qing, Wang Jian

渤龙湖总部基地二区 ●
Section 2 of Bolonghu Headquarter Base
项目地点 / Location：天津 / Tianjin
设计时间 / Design Period：2009.9—2011.12
施工时间 / Construction Period：2009.12—2019.12
用地面积 / Site Area：192,726 m²
建筑面积 / Floor Area：297,111 m²
建筑设计 / Architecture：李兴钢、邱涧冰、李宁、薛从清、闫昱、肖育智、邢迪、唐勇 / Li Xinggang, Qiu Jianbing, Li Ning, Xue Congqing, Yan Yu, Xiao Yuzhi, Xing Di, Tang Yong
结构设计 / Structure：张付奎 / Zhang Fukui
机电设计 / Engineering：宋国清、高振渊、宋孝春、邬可文、胡桃 / Song Guoqing, Gao Zhenyuan, Song Xiaochun, Wu Kewen, Hu Tao
总图设计 / Master Plan：吴耀懿 / Wu Yaoyi

绩溪博物馆 ●
Jixi Museum
项目地点 / Location：安徽绩溪 / Jixi, Anhui
设计时间 / Design Period：2009.11—2010.12
施工时间 / Construction Period：2010.12—2013.11
用地面积 / Site Area：9,500 m²
建筑面积 / Floor Area：10,003 m²
建筑设计 / Architecture：李兴钢、张音玄、张哲、邢迪、张一婷、易灵洁、钟曼琳 / Li Xinggang, Zhang Yinxuan, Zhang Zhe, Xing Di, Zhang Yiting, Yi Lingjie, Zhong Manlin
结构设计 / Structure：王立波、杨威、梁伟 / Wang Libo, Yang Wei, Liang Wei
机电设计 / Engineering：何猛、李京沙、张千、李俊民、丁志强 / He Meng, Li Jingsha, Zhang Qian, Li Junmin, Ding Zhiqiang
景观设计 / Landscape：李力、于超 / Li Li, Yu Chao

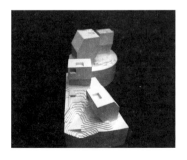

鄂尔多斯 20+10 项目 D4、P19 地块
D4, P19 of Ordos 20+10 Project
项目地点 / Location：内蒙古鄂尔多斯 / Ordos, Inner Mongolia
设计时间 / Design Period：2010.2—2011.6
用地面积 / Site Area：10,440 m²
建筑面积 / Floor Area：33,053 m²
建筑设计 / Architecture：李兴钢、谭泽阳、邢迪、唐勇 / Li Xinggang, Tan Zeyang, Xing Di, Tang Yong
结构设计 / Structure：王立波、杨威、梁伟 / Wang Libo, Yang Wei, Liang Wei

MAX LAB IV（竞赛方案，与瑞典 FOJAB 合作）
MAX LAB IV (Competition, cooperated with FOJAB, Sweden)
项目地点 / Location：瑞典隆德 / Lund, Sweden
设计时间 / Design Period：2010.7—2010.10
用地面积 / Site Area：197,687 m²
建筑面积 / Floor Area：37,860 m²
建筑设计 / Architecture：李兴钢、张玉婷、易灵洁、Ulf Kadefors、Rachelle Astrand / Li Xinggang, Zhang Yuting, Yi Lingjie, Ulf Kadefors, Rachelle Astrand

元上都遗址工作站 ●
Visitor's Center for Site of Xanadu
项目地点 / Location：内蒙古正蓝旗 / Zhenglanqi, Inner Mongolia
设计时间 / Design Period：2010.8—2011.5
施工时间 / Construction Period：2011.5—2011.8
用地面积 / Site Area：16,663 m²
建筑面积 / Floor Area：410 m²
建筑设计 / Architecture：李兴钢、邱涧冰、易灵洁、孙鹏、张玉婷、赵小雨 / Li Xinggang, Qiu Jianbing, Yi Lingjie, Sun Peng, Zhang Yuting, Zhao Xiaoyu
结构设计 / Structure：高银鹰 / Gao Yinying
机电设计 / Engineering：刘海、宋孝春、李超英、甄毅 / Liu Hai, Song Xiaochun, Li Chaoying, Zhen Yi
景观设计 / Landscape：余晓东 / Yu Xiaodong

西柏坡华润希望小镇 ●
Xibaipo China Resources Hope Town
项目地点 / Location：河北平山 / Pingshan, Hebei
设计时间 / Design Period：2010.9—2011.3
施工时间 / Construction Period：2011.3—2012.8
用地面积 / Site Area：153,800 m²
建筑面积 / Floor Area：53,100 m²
建筑设计 / Architecture：李兴钢、邱涧冰、梁旭、张一婷、马津、赵小雨、唐勇、李喆 / Li Xinggang, Qiu Jianbing, Liang Xu, Zhang Yiting, Ma Jin, Zhao Xiaoyu, Tang Yong, Li Zhe
结构设计 / Structure：毕磊、何羽、何喜明 / Bi Lei, He Yu, He Ximing
机电设计 / Engineering：石小飞、王浩然、林佳、金健 / Shi Xiaofei, Wang Haoran, Lin Jia, Jin Jian
总图设计 / Master Plan：吴耀懿 / Wu Yaoyi

北京 CBD 核心区（城市设计）↖
Core Area of Beijing CBD (Urban Design)
项目地点 / Location：北京 / Beijing
设计时间 / Design Period：2010.9—2010.11
用地面积 / Site Area：246,720 m²
建筑面积 / Floor Area：2,000,000 m²
建筑设计 / Architecture：李兴钢、张玉婷、李喆、易灵洁 / Li Xinggang, Zhang Yuting, Li Zhe, Yi Lingjie

北京通州新城运河核心区 VII09-14 用地 ↖
Canal Core Area VII09-14, Tongzhou New Town
项目地点 / Location：北京 / Beijing
设计时间 / Design Period：2010.10
用地面积 / Site Area：83,453 m²
建筑面积 / Floor Area：700,000 m²
建筑设计 / Architecture：李兴钢、张玉婷、李喆 / Li Xinggang, Zhang Yuting, Li Zhe

中国国学中心（竞赛方案）↖
China Sinology Centre (Competition)
项目地点 / Location：北京 / Beijing
设计时间 / Design Period：2010.12—2011.11
用地面积 / Site Area：20,500 m²
建筑面积 / Floor Area：65,928 m²
建筑设计 / Architecture：李兴钢、张音玄、张哲、张玉婷、闫昱、李宁、邢迪、李喆、唐勇、梁旭、易灵洁 / Li Xinggang, Zhang Yinxuan, Zhang Zhe, Zhang Yuting, Yan Yu, Li Ning, Xing Di, Li Zhe, Tang Yong, Liang Xu, Yi Lingjie

天津大学新校区综合体育馆●
Gymnasium of the New Campus of Tianjin University
项目地点 / Location：天津 / Tianjin
设计时间 / Design Period：2011.2—2013.8
施工时间 / Construction Period：2013.8—2015.11
用地面积 / Site Area：33,950 m²
建筑面积 / Floor Area：18,798 m²
建筑设计 / Architecture：李兴钢、张音玄、闫昱、易灵洁、梁旭 / Li Xinggang, Zhang Yinxuan, Yan Yu, Yi Lingjie, Liang Xu
结构设计 / Structure：任庆英、张付奎、李森 / Ren Qingying, Zhang Fukui, Li Sen
机电设计 / Engineering：赵昕、李建业、王微微、唐艳滨、王旭 / Zhao Xin, Li Jianye, Wang Weiwei, Tang Yanbin, Wang Xu

北京朝阳区生活垃圾综合处理厂焚烧中心●
Waste Treatment Center in Chaoyang
项目地点 / Location：北京 / Beijing
设计时间 / Design Period：2011.2—2014.7
施工时间 / Construction Period：2013.12—2016.10
用地面积 / Site Area：49,000 m²
建筑面积 / Floor Area：36,000 m²
建筑设计 / Architecture：李兴钢、张音玄、梁旭 / Li Xinggang, Zhang Yinxuan, Liang Xu
结构设计 / Structure：中国城市建设研究院有限公司 / China Urban Construction Design & Research Institute Co., Ltd.
机电设计 / Engineering：中国城市建设研究院有限公司 / China Urban Construction Design & Research Institute Co., Ltd.

中国建筑设计院新楼（创新科研示范中心）（竞赛方案）↖
 BIM Ecology Office Building of China Architecture Design & Research Group (Competition)
项目地点 / Location：北京 / Beijing
设计时间 / Design Period：2011.4—2011.5
用地面积 / Site Area：4,114 m²
建筑面积 / Floor Area：33,108 m²
建筑设计 / Architecture：李兴钢、张音玄、张哲、闫昱、唐勇、李喆 / Li Xinggang, Zhang Yinxuan, Zhang Zhe, Yan Yu, Tang Yong, Li Zhe

中国驻西班牙大使馆改造 ↲
Renovation of the Office Building of Chinese Embassy in Spain
项目地点 / Location：西班牙马德里 / Madrid, Spain
设计时间 / Design Period：2011.6—2013.5
施工时间 / Construction Period：2013.5—
用地面积 / Site Area：2,400 m²
建筑面积 / Floor Area：3,355 m²
建筑设计 / Architecture：李兴钢、邱涧冰、张音玄、谭泽阳、易灵洁、梁旭、姜汶林、朱伶俐、侯新觉 / Li Xinggang, Qiu Jianbing, Zhang Yinxuan, Tan Zeyang, Yi Lingjie, Liang Xu, Jiang Wenlin, Zhu Lingli, Hou Xinjue
结构设计 / Structure：张付奎、刘文斑、刘福、杨杰、李硕 / Zhang Fukui, Liu Wenting, Liu Fu, Yang Jie, Li Shuo
机电设计 / Engineering：申静、李建业、王微微、祝秀娟、林佳、王旭、张辉 / Shen Jing, Li Jianye, Wang Weiwei, Zhu Xiujuan, Lin Jia, Wang Xu, Zhang Hui
照明顾问 / Lightning：郑见伟 / Zheng Jianwei

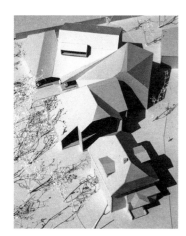

中国驻爱沙尼亚大使馆办公楼 ↖
Office Building of Chinese Embassy in Estonia
项目地点 / Location：爱沙尼亚塔林 / Tallinn, Estonia
设计时间 / Design Period：2011.6—
用地面积 / Site Area：7,270 m²
建筑面积 / Floor Area：5,960 m²
建筑设计 / Architecture：李兴钢、张音玄、张哲、李喆、姜汶林 / Li Xinggang, Zhang Yinxuan, Zhang Zhe, Li Zhe, Jiang Wenlin

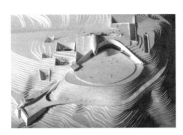

郧阳博物馆 ⬧
Yunyang Museum
项目地点 / Location：湖北郧县 / Yun County, Hubei
设计时间 / Design Period：2011.7—2014.10
用地面积 / Site Area：26,646 m²
建筑面积 / Floor Area：10,428 m²
建筑设计 / Architecture：李兴钢、张音玄、易灵洁、邢迪、刘振 / Li Xinggang, Zhang Yinxuan, Yi Lingjie, Xing Di, Liu Zhen
结构设计 / Structure：鲁昂、王磊 / Lu Ang, Wang Lei
机电设计 / Engineering：刘海、宋孝春、李超英、甄毅 / Liu Hai, Song Xiaochun, Li Chaoying, Zhen Yi
总图设计 / Master Plan：吴耀懿 / Wu Yaoyi

新疆第十三届全国冬运会冰上项目场馆（竞赛方案）⬧
Xinjiang 13th National Winter Games Stadium (Competition)
项目地点 / Location：新疆乌鲁木齐 / Urumqi, Xinjiang
设计时间 / Design Period：2012.3—2012.4
用地面积 / Site Area：366,850 m²
建筑面积 / Floor Area：83,415 m²
建筑设计 / Architecture：李兴钢、张音玄、邱涧冰、张玉婷、赵小雨、梁旭、孙鹏、邢迪、李宁 / Li Xinggang, Zhang Yinxuan, Qiu Jianbing, Zhang Yuting, Zhao Xiaoyu, Liang Xu, Sun Peng, Xing Di, Li Ning

海南蓝海裕华大酒店 ☽
Hainan Luxury Blue Horizon Hotel
项目地点 / Location：海南澄迈 / Chengmai, Hainan
设计时间 / Design Period：2012.3—2015.9
施工时间 / Construction Period：2013.1—
用地面积 / Site Area：29,375 m²
建筑面积 / Floor Area：72,373 m²
建筑设计 / Architecture：李兴钢、邱涧冰、张哲、朱伶俐、孙鹏、李欢、侯新觉、刘紫琪、刘振、闫昱、张司腾 / Li Xinggang, Qiu Jianbing, Zhang Zhe, Zhu Lingli, Sun Peng, Li Huan, Hou Xinjue, Liu Ziqi, Liu Zhen, Yan Yu, Zhang Siteng
结构设计 / Structure：鲁昂 / Lu Ang
机电设计 / Engineering：赵昕、康国青、马霄鹏、林佳 / Zhao Xin, Kang Guoqing, Ma Xiaopeng, Lin Jia
总图设计 / Master Plan：高治、高伟 / Gao Zhi, Gao Wei
照明顾问 / Lighting：郑见伟 / Zheng Jianwei

鸟巢文化中心 ●
Bird's Nest Cultural Center
项目地点 / Location：北京 / Beijing
设计时间 / Design Period：2012.10—2013.1
施工时间 / Construction Period：2013.5—2015.3
建筑面积 / Floor Area：7,598 m²
建筑设计 / Architecture：李兴钢、谭泽阳、张玉婷、张司腾、唐勇 / Li Xinggang, Tan Zeyang, Zhang Yuting, Zhang Siteng, Tang Yong
结构设计 / Structure：尤天直、王大庆 / You Tianzhi, Wang Daqing
机电设计 / Engineering：郭汝艳、张庆康、曹磊、刘征峥 / Guo Ruyan, Zhang Qingkang, Cao Lei, Liu Zhengzheng

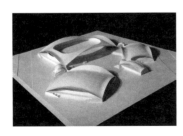

吕梁体育中心 ☽
Lvliang Sports Centre
项目地点 / Location：山西吕梁 / Lvliang, Shanxi
设计时间 / Design Period：2012.10—
施工时间 / Construction Period：2013.7—
用地面积 / Site Area：201,451 m²
建筑面积 / Floor Area：81,208 m²
建筑设计 / Architecture：李兴钢、谭泽阳、张音玄、邢迪、李喆、唐勇、张哲、张司腾、董秀芳、朱伶俐、周威、李昀倩 / Li Xinggang, Tan Zeyang, Zhang Yinxuan, Xing Di, Li Zhe, Tang Yong, Zhang Zhe, Zhang Siteng, Dong Xiufang, Zhu Lingli, Zhou Wei, Li Yunqian
结构设计 / Structure：任庆英、刘文珽、王奇 / Ren Qingying, Liu Wenting, Wang Qi
机电设计 / Engineering：王则慧、李峰、朱永智、王思乡、王陈栋 / Wang Zehui, Li Feng, Zhu Yongzhi, Wang Sixiang, Wang Chendong
景观设计 / Landscape：张景华、刘环 / Zhang Jinghua, Liu Huan

玉环博物馆和图书馆 ☽
Yuhuan Museum & Library
项目地点 / Location：浙江玉环 / Yuhuan, Zhejiang
设计时间 / Design Period：2013.3—2014.10
施工时间 / Construction Period：2015.11—
用地面积 / Site Area：26,646 m²
建筑面积 / Floor Area：10,428 m²
建筑设计 / Architecture：李兴钢、张音玄、闫昱、张司腾 / Li Xinggang, Zhang Yinxuan, Yan Yu, Zhang Siteng
结构设计 / Structure：任庆英、刘文珽、张雄迪、李森 / Ren Qingying, Liu Wenting, Zhang Xiongdi, Li Sen
机电设计 / Engineering：袁乃荣、范改娜、韦航、姜红、贾京花、王莉、李甲 / Yuan Nairong, Fan Gaina, Wei Hang, Jiang Hong, Jia Jinghua, Wang Li, Li Jia
总图设计 / Master Plan：朱秀丽 / Zhu Xiuli
室内设计 / Interior：曹阳、马萌雪、李晓菲 / Cao Yang, Ma Mengxue, Li Xiaofei
景观设计 / Landscape：杨陈 / Yang Chen
照明顾问 / Lighting：张昕照明设计工作室 / Zhang Xin Lighting Studio

泉州当代艺术馆旧馆一期 ●
Quanzhou Contemporary Art Museum
项目地点 / Location：福建泉州 / Quanzhou, Fujian
设计时间 / Design Period：2013.9—2014.2
施工时间 / Construction Period：2013.12—
用地面积 / Site Area：35,613 m²
建筑面积 / Floor Area：40,000 m²
建筑设计 / Architecture：李兴钢、谭泽阳、张玉婷、姜汶林 / Li Xinggang, Tan Zeyang, Zhang Yuting, Jiang Wenlin
结构设计 / Structure：廖扬、张文英 / Liao Yang, Zhang Wenying
机电设计 / Engineering：朱跃云、曹磊 / Zhu Yueyun, Cao Lei

通辽美术馆和蒙古族服饰博物馆 ●
Tongliao Art Gallery & Mongolian Costume Museum
项目地点 / Location：内蒙古通辽 / Tongliao, Inner Mongolia
设计时间 / Design Period：2014.3—2014.5
施工时间 / Construction Period：2014.5—2017.9
用地面积 / Site Area：3,788m²、8,366m²
建筑面积 / Floor Area：2,142m²、4,983m²
建筑设计 / Architecture：李兴钢、张音玄、梁旭、杜捷、邢迪、王辰、祃冀然、刘津津 / Li Xinggang, Zhang Yinxuan, Liang Xu, Du Jie, Xing Di, Wang Chen, Ma Jiran, Liu Jinjin
结构设计 / Structure：刘建涛 / Liu Jiantao
机电设计 / Engineering：张丽、沈辉、贺琳 / Zhang Li, Shen Hui, He Lin
总图设计 / Master Plan：吴耀懿 / Wu Yaoyi

南京安品园舍 ●
Anpin Garden Houses in Nanjing
项目地点 / Location：江苏南京 / Nanjing, Jiangsu
设计时间 / Design Period：2014.3—2016.12
施工时间 / Construction Period：2015.9—2019.8
用地面积 / Site Area：27,325 m²
建筑面积 / Floor Area：61,224 m²
建筑设计 / Architecture：李兴钢、张音玄、张哲、刘振、易灵洁、邱涧冰、闫昱、王汉、王子昂 / Li Xinggang, Zhang Yinxuan, Zhang Zhe, Liu Zhen, Yi Lingjie, Qiu Jianbing, Yan Yu, Wang Han, Wang Zi'ang
结构设计 / Structure：南京长江都市建筑设计股份有限公司 / Nanjing Yangtze River Urban Architectural Design Co., Ltd.
机电设计 / Engineering：南京长江都市建筑设计股份有限公司 / Nanjing Yangtze River Urban Architectural Design Co., Ltd.

济南小清河工作坊垂钓中心 ◟
Fishing Center of Xiaoqing River in Jinan
项目地点 / Location： 山东济南 / Jinan, Shandong
设计时间 / Design Period：2014.8—2015.1
用地面积 / Site Area： 20,842 m^2
建筑面积 / Floor Area： 2,262 m^2
建筑设计 / Architecture： 李兴钢、张音玄、张司腾、姜汶林 / Li Xinggang, Zhang Yinxuan, Zhang Siteng, Jiang Wenlin

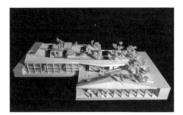

深圳万科云城北绿廊 05-03 地块中区 ◟
Block 05-03 of Vanke Cloud City in Shenzhen
项目地点 / Location： 广东深圳 / Shenzhen, Guangdong
设计时间 / Design Period：2014.9—2016.8
用地面积 / Site Area： 27,512 m^2
建筑面积 / Floor Area： 41,146 m^2
建筑设计 / Architecture： 李兴钢、张音玄、梁旭、张哲 / Li Xinggang, Zhang Yinxuan, Liang Xu, Zhang Zhe

"聚落"卡座 ●
"Set-all"
项目地点 / Location： 上海 / Shanghai
设计时间 / Design Period：2015.3
施工时间 / Construction Period：2015.4
建筑设计 / Architecture： 李兴钢、张玉婷 / Li Xinggang, Zhang Yuting

国家体育场 2015 年世锦赛注册中心 ●
Registration Center of IAAF World Championships, Beijing 2015
项目地点 / Location： 北京 / Beijing
设计时间 / Design Period：2015.3—2015.5
施工时间 / Construction Period：2015.4—2015.8
用地面积 / Site Area： 1,869 m^2
建筑面积 / Floor Area： 1,417 m^2
建筑设计 / Architecture： 李兴钢、谭泽阳、张司腾 / Li Xinggang, Tan Zeyang, Zhang Siteng
结构设计 / Structure： 杨松霖、曾金盛 / Yang Songlin, Zeng Jinsheng
机电设计 / Engineering： 关维、张昕、王京生 / Guan Wei, Zhang Xin, Wang Jingsheng

瞬时桃花源（已拆除）●
Instantaneous Peach Garden (Demolished)
项目地点 / Location： 江苏南京 / Nanjing, Jiangsu
设计时间 / Design Period：2015.5
施工时间 / Construction Period：2015.7
建筑面积 / Floor Area： 176 m^2
建筑设计 / Architecture： 李兴钢、张玉婷、姜汶林 / Li Xinggang, Zhang Yuting, Jiang Wenlin

中国驻新西兰大使馆 ◟
Chinese Embassy in New Zealand
项目地点 / Location： 新西兰惠灵顿 / Wellington, New Zealand
设计时间 / Design Period：2015.9—
用地面积 / Site Area： 10,244 m^2
建筑面积 / Floor Area： 12,077 m^2
建筑设计 / Architecture： 李兴钢、张音玄、邱涧冰、易灵洁、姜汶林 / Li Xinggang, Zhang Yinxuan, Qiu Jianbing, Yi Lingjie, Jiang Wenlin

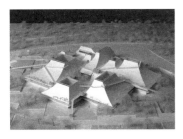

延庆世园会国际馆（竞赛方案）🔺
International Pavilion of International Horticultural Exhibition 2019 in Yanqing (Competition)
项目地点 / Location：北京 / Beijing
设计时间 / Design Period：2015.10—2016.2
用地面积 / Site Area：40,303 m²
建筑面积 / Floor Area：22,890 m²
建筑设计 / Architecture：李兴钢、张音玄、张哲、刘振、张司腾、冯方娜、张耀飞 / Li Xinggang, Zhang Yinxuan, Zhang Zhe, Liu Zhen, Zhang Siteng, Feng Fangna, Zhang Yaofei

厦门音乐中心（竞赛方案）🔺
Xiamen Music Center (Competition)
项目地点 / Location：福建厦门 / Xiamen, Fujian
设计时间 / Design Period：2015.10—2015.12
用地面积 / Site Area：46,977 m²
建筑面积 / Floor Area：38,402 m²
建筑设计 / Architecture：李兴钢、张玉婷、闫昱、姜汶林、朱伶俐、张司腾、王汉 / Li Xinggang, Zhang Yuting, Yan Yu, Jiang Wenlin, Zhu Lingli, Zhang Siteng, Wang Han

华夏幸福幼儿园 🔺
China Fortune Land Kindergarten
项目地点 / Location：河北香河 / Xianghe, Hebei
设计时间 / Design Period：2015.11—2016.6
用地面积 / Site Area：5,262 m²
建筑面积 / Floor Area：2,577 m²
建筑设计 / Architecture：李兴钢、邱涧冰、张玉婷、韩智华、姜汶林、朱伶俐、林景怡、刘苗苗、马莹、关晓旭 / Li Xinggang, Qiu Jianbing, Zhang Yuting, Han Zhihua, Jiang Wenlin, Zhu Lingli, Lin Jingyi, Liu Miaomiao, Ma Ying, Guan Xiaoxu
结构设计 / Structure：许庆、刘家名 / Xu Qing, Liu Jiaming
机电设计 / Engineering：刘云朔、徐阳、张恩茂 / Liu Yunshuo, Xu Yang, Zhang Enmao
照明顾问 / Lighting：张昕照明设计工作室 / Zhang Xin Lighting Studio

"仓阁"——首钢工舍智选假日酒店 ●
"Silo Pavilion", Holiday Inn Express Beijing Shougang
项目地点 / Location：北京 / Beijing
设计时间 / Design Period：2015.12—2016.8
施工时间 / Construction Period：2016.5—2018.5
用地面积 / Site Area：2,100 m²
建筑面积 / Floor Area：9,890 m²
建筑设计 / Architecture：李兴钢、景泉、黎靓、郑旭航、涂嘉欢 / Li Xinggang, Jing Quan, Li Liang, Zheng Xuhang, Tu Jiahuan
结构设计 / Structure：王树乐、郭俊杰 / Wang Shule, Guo Junjie
机电设计 / Engineering：申静、郝洁、祝秀娟、高学文 / Shen Jing, Hao Jie, Zhu Xiujuan, Gao Xuewen
室内设计 / Interior：曹阳 / Cao Yang

重庆两江体艺中心 🔺
Liangjiang Sports and Arts Center in Chongqing
项目地点 / Location：重庆 / Chongqing
设计时间 / Design Period：2016.1—2016.3
用地面积 / Site Area：72,385 m²
建筑面积 / Floor Area：175,000 m²
建筑设计 / Architecture：李兴钢、张哲、张司腾、李欢、闫昱 / Li Xinggang, Zhang Zhe, Zhang Siteng, Li Huan, Yan Yu

怀柔水长城书院
Great Wall Academy in Huairou
项目地点 / Location：北京 / Beijing
设计时间 / Design Period：2016.2—2016.10
用地面积 / Site Area：29,255 m²
建筑面积 / Floor Area：28,135 m²
建筑设计 / Architecture：李兴钢、张音玄、谭泽阳、梁旭、李欢、陆少波 / Li Xinggang, Zhang Yinxuan, Tan Zeyang, Liang Xu, Li Huan, Lu Shaobo
结构设计 / Structure：杨婷 / Yang Ting
机电设计 / Engineering：夏树威、刘洞阳、唐艳滨、李战赠、祁桐 / Xia Shuwei, Liu Dongyang, Tang Yanbin, Li Zhanzeng, Qi Tong
总图设计 / Master Plan：高治 / Gao Zhi
景观设计 / Landscape：关午军 / Guan Wujun
室内设计 / Interior：王强、纪岩 / Wang Qiang, Ji Yan

上海万科雅宾利小学和幼儿园
Yabinli School and Kindergarten of Vanke in Shanghai
项目地点 / Location：上海 / Shanghai
设计时间 / Design Period：2016.3—
用地面积 / Site Area：15,189 m²
建筑面积 / Floor Area：23,484 m²
建筑设计 / Architecture：李兴钢、刘振、张音玄、张哲、冯方娜、陆婧瑶、赵戈、朱伶俐 / Li Xinggang, Liu Zhen, Zhang Yinxuan, Zhang Zhe, Feng Fangna, Lu Jingyao, Zhao Ge, Zhu Lingli

成都体育中心改造
Renovation of Chengdu Sports Center
项目地点 / Location：四川成都 / Chengdu, Sichuan
设计时间 / Design Period：2016.3—2017.4
用地面积 / Site Area：36,000 m²
建筑面积 / Floor Area：53,589 m²
建筑设计 / Architecture：李兴钢、谭泽阳、闫昱、张司腾、李欢 / Li Xinggang, Tan Zeyang, Yan Yu, Zhang Siteng, Li Huan

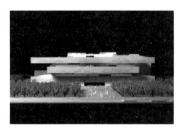

上海博物馆东馆（竞赛方案）
Shanghai East Museum (Competition)
项目地点 / Location：上海 / Shanghai
设计时间 / Design Period：2016.9—2017.1
用地面积 / Site Area：46,001 m²
建筑面积 / Floor Area：110,638 m²
建筑设计 / Architecture：李兴钢、张音玄、张玉婷、姜汶林、王汉、刘紫骐、谭舟、李慧、张司腾 / Li Xinggang, Zhang Yinxuan, Zhang Yuting, Jiang Wenlin, Wang Han, Liu Ziqi, Tan Zhou, Li Hui, Zhang Siteng
结构设计 / Structure：范重、刘学林 / Fan Zhong, Liu Xuelin

叠合院——护国寺西巷 37 号院改造
The Stack Courtyard House
项目地点 / Location：北京 / Beijing
设计时间 / Design Period：2016.11—2017.3
用地面积 / Site Area：216 m²
建筑面积 / Floor Area：132 m²
建筑设计 / Architecture：李兴钢、谭泽阳、朱伶俐、张捍平、侯新觉、王汉 / Li Xinggang, Tan Zeyang, Zhu Lingli, Zhang Hanping, Hou Xinjue, Wang Han

楼纳大冲组民宿改造
Home Stay in Louna
项目地点 / Location：贵州兴义 / Xingyi, Guizhou
设计时间 / Design Period：2016.12—2017.3
建筑面积 / Floor Area：2,452 m²
建筑设计 / Architecture：李兴钢、张哲、梁旭、刘振、李欢 / Li Xinggang, Zhang Zhe, Liang Xu, Liu Zhen, Li Hua

楼纳咖啡馆改造
Cafe in Louna
项目地点 / Location: 贵州兴义 / Xingyi, Guizhou
设计时间 / Design Period: 2016.11—2017.3
用地面积 / Site Area: 559 m²
建筑面积 / Floor Area: 461 m²
建筑设计 / Architecture: 李兴钢、易灵洁、王继飞、Elizaveta/ Li Xinggang, Yi Lingjie, Wang Jifei, Elizaveta

楼纳露营服务中心 ●
Camping Service Center in Louna
项目地点 / Location: 贵州兴义 / Xingyi, Guizhou
设计时间 / Design Period: 2016.11—2017.7
施工时间 / Construction Period: 2016.12—2017.10
用地面积 / Site Area: 306 m²
建筑面积 / Floor Area: 1,863 m²
建筑设计 / Architecture: 李兴钢、谭泽阳、陆少波、侯新觉、刘智奇、师宏刚 / Li Xinggang, Tan Zeyang, Lu Shaobo, Hou Xinjue, Liu Zhiqi, Shi Honggang
结构设计 / Structure: 赵广海 / Zhao Guanghai
机电设计 / Engineering: 许艳、程斌 / Xu Yan, Cheng Bin

"微缩北京"——大院胡同 28 号院改造 ●
"Miniature Beijing", Renovation of No. 28 Dayuan Hutong
项目地点 / Location: 北京 / Beijing
设计时间 / Design Period: 2016.11—2017.4
施工时间 / Construction Period: 2017.4—2017.9
用地面积 / Site Area: 262 m²
建筑面积 / Floor Area: 214 m²
建筑设计 / Architecture: 李兴钢、谭泽阳、朱伶俐、侯新觉、张捍平、谭舟 / Li Xinggang, Tan Zeyang, Zhu Lingli, Hou Xinjue, Zhang Hanping, Tan Zhou
结构设计 / Structure: 王树乐、李博 / Wang Shule, Li Bo
机电设计 / Engineering: 刘洞阳、董元君、常立强 / Liu Dongyang, Dong Yuanjun, Chang Liqiang

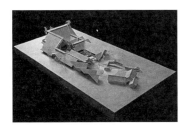

安仁大匠之门文化中心 ●
Grand Carpenter's Gate Cultural Center in Anren
项目地点 / Location: 四川大邑 / Dayi, Sichuan
设计时间 / Design Period: 2017.3—
用地面积 / Site Area: 12,806 m²
建筑面积 / Floor Area: 7,441 m²
建筑设计 / Architecture: 李兴钢、刘振、陈译民、王汉、郭文嘉 / Li Xinggang, Liu Zhen, Chen Yimin, Wang Han, Guo Wenjia
结构设计 / Structure: 何相宇、陈晓晴 / He Xiangyu, Chen Xiaoqing
机电设计 / Engineering: 刘洞阳、朱琳、唐艳滨、常立强、蒲域 / Liu Dongyang, Zhu Lin, Tang Yanbin, Chang Liqiang, Pu Yu
总图设计 / Master Plan: 高治 / Gao Zhi
室内设计 / Interior: 中设筑邦建筑设计研究院张明杰设计工作室 / Zhang Mingjie Studio, LBY Architectural Design Co.,Ltd.
照明顾问 / Lighting: 张昕照明设计工作室 / Zhang Xin Lightning Studio

中国驻马耳他使馆
Chinese embassy in Republic of Malta
项目地点 / Location: 马耳他瓦莱塔 / Valletta, Republic of Malta
设计时间 / Design Period: 2017.4—
用地面积 / Site Area: 19,115 m²
建筑面积 / Floor Area: 10,617 m²
建筑设计 / Architecture: 李兴钢、张音玄、魏鸣宇 / Li Xinggang, Zhang Yinxuan, Wei Mingyu

屺园——延庆园艺小镇文创中心 ●
Mountain Garden, Cultural Center of Horticulture Village in Yanqing
项目地点 / Location：北京 / Beijing
设计时间 / Design Period：2017.6—2018.10
施工时间 / Construction Period：2018.10—2019.4
用地面积 / Site Area：2,175 m²
建筑面积 / Floor Area：2,310 m²
建筑设计 / Architecture：李兴钢、谭泽阳、姜汶林、袁智敏 / Li Xinggang, Tan Zeyang, Jiang Wenlin, Yuan Zhimin
结构设计 / Structure：刘文珽、杨松霖 / Liu Wenting, Yang Songlin
机电设计 / Engineering：申静、郝洁、祝秀娟、王旭 / Shen Jing, Hao Jie, Zhu Xiujuan, Wang Xu
照明顾问 / Lighting：张昕照明设计工作室 / Zhang Xin Lighting Studio

崇礼太子城雪花小镇 ◗
Snowflake Town of Prince City in Chongli
项目地点 / Location：河北张家口 / Zhangjiakou, Hebei
设计时间 / Design Period：2017.11—2019.4
施工时间 / Construction Period：2018.8—
用地面积 / Site Area：165,504 m²
建筑面积 / Floor Area：333,399 m²
建筑设计 / Architecture：李兴钢、易灵洁、王汉、沈周娅、李欢、谭舟、陆婧瑶、郭永健、谭泽阳、张司腾、孙知行、郭文嘉、侯新觉、梁艺晓、李慧、李东哲、王凌云、涂嘉欢、潘悦 / Li Xinggang, Yi Lingjie, Wang Han, Shen Zhouya,Li Huan, Tan Zhou, Lu Jingyao, Guo Yongjian, Tan Zeyang, Zhang Siteng, Sun Zhixing, Guo Wenjia, Hou Xinjue, Liang Yixiao, Li Hui, Li Dongzhe, Wang Lingyun, Tu Jiahuan, Pan Yue
结构设计 / Structure：潘敏华、董越 / Pan Minhua, Dong Yue
机电设计 / Engineering：唐致文、安岩、李梅、李嘉、康向东、刘畅 / Tang Zhiwen, An Yan, Li Mei, Li Jia, Kang Xiangdong, Liu Chang
总图设计 / Master Plan：齐海娟 / Qi Haijuan

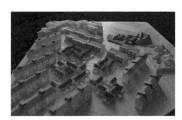

安仁金井小镇 ◗
Jinjing Community in Anren
项目地点 / Location：四川大邑 / Dayi, Sichuan
设计时间 / Design Period：2018.1—
用地面积 / Site Area：64,265 m²
建筑面积 / Floor Area：88,052 m²
建筑设计 / Architecture：李兴钢、刘振、孙知行、于安然、侯新觉、梁艺晓、姜汶林、李昀倩 / Li Xinggang, Liu Zhen, Sun Zhixing, Yu Anran, Hou Xinjue, Liang Yixiao, Jiang Wenlin, Li Yunqian
结构设计 / Structure：中国建筑技术集团有限公司 / China Building Technique Group Co.,Ltd.
机电设计 / Engineering：中国建筑技术集团有限公司 / China Building Technique Group Co.,Ltd.
景观设计 / Landscape：关午军、路璐、冯然 / Guan Wujun, Lu Lu, Feng Ran

壁园 ●
Biyuan Garden
项目地点 / Location：北京 / Beijing
设计时间 / Design Period：2018.4—2018.7
施工时间 / Construction Period：2018.7—2019.8
用地面积 / Site Area：38 m²
建筑面积 / Floor Area：6 m²
建筑设计 / Architecture：李兴钢、侯新觉、孙宇 / Li Xinggang, Hou Xinjue, Sun Yu

斗园——厦门十九集美 C 段 5 号地主题书店 ◣
Douyuan Garden, A Bookstore in Xiamen
项目地点 / Location：福建厦门 / Xiamen, Fujian
设计时间 / Design Period：2018.5—2019.5
用地面积 / Site Area：7,079 m²
建筑面积 / Floor Area：5,261 m²
建筑设计 / Architecture：李兴钢、姜汶林 / Li Xinggang, Jiang Wenlin

舞雩亭——衢州生态公厕 ▸
Wuyu Pavilion, Toilet in Quzhou
项目地点 / Location：浙江衢州 / Quzhou, Zhejiang
设计时间 / Design Period：2018.6—2020.1
用地面积 / Site Area：3,000 m²
建筑面积 / Floor Area：300 m²
建筑设计 / Architecture：李兴钢、易灵洁、袁智敏、Luis Michal / Li Xinggang, Yi Lingjie, Yuan Zhimin, Luis Michal
结构设计 / Structure：董越 / Dong Yue
机电设计 / Engineering：沈晨、李嘉、刘畅 / Shen Chen, Li Jia, Liu Chang

廖维公馆改造暨安仁古镇游客服务中心 ☽
Renovation of Liaowei Mansion
项目地点 / Location：四川大邑 / Dayi, Sichuan
设计时间 / Design Period：2018.10—2019.10
施工时间 / Construction Period：2019.6—
用地面积 / Site Area：16,064 m²
建筑面积 / Floor Area：3,027 m²
建筑设计 / Architecture：李兴钢、刘振、郭文嘉、于安然 / Li Xinggang, Liu Zhen, Guo Wenjia, Yu Anran
结构设计 / Structure：何相宇、陈晓晴 / He Xiangyu, Chen Xiaoqing
机电设计 / Engineering：刘洞阳、汪春华、蒲域 / Liu Dongyang, Wang Chunhua, Pu Yu
总图设计 / Master Plan：高治 / Gao Zhi
景观设计 / Landscape：关午军、路璐、冯然 / Guan Wujun, Lu Lu, Feng Ran
室内设计 / Interior：曹阳、范小胜、张然 / Cao Yang, Fan Xiaosheng, Zhang Ran

崇台——北京冬奥会张家口赛区奥运展示中心 ☽
Chongtai, Winter Olympic Exhibition Center in Chongli
项目地点 / Location：河北张家口 / Zhangjiakou, Hebei
设计时间 / Design Period：2018.12—2019.9
施工时间 / Construction Period：2020.3—
用地面积 / Site Area：6,142 m²
建筑面积 / Floor Area：12,000 m²
建筑设计 / Architecture：李兴钢、易灵洁、谭泽阳、袁智敏 / Li Xinggang, Yi Lingjie, Tan Zeyang, Yuan Zhimin
结构设计 / Structure：孙海林、孙庆唐 / Sun Hailin, Sun Qingtang
机电设计 / Engineering：张庆康、李斌、全巍、于征 / Zhang Qingkang, Li Bin, Quan Wei, Yu Zheng
总图设计 / Master Plan：齐海娟 / Qi Haijuan
室内设计 / Interior：王国彬、余深宏 / Wang Guobin, Yu Shenhong

东城区工人文化宫改造 ▸
Renovation of Workers Club in Dongcheng
项目地点 / Location：北京 / Beijing
设计时间 / Design Period：2019.1—
用地面积 / Site Area：3,174 m²
建筑面积 / Floor Area：4,345 m²
建筑设计 / Architecture：李兴钢、郑世伟、罗云、胥悦 / Li Xinggang, Zheng Shiwei, Luo Yun, Xu Yue

中共洛阳组诞生地纪念馆及周边城市设计 ▸
CPC Luoyang Branch Birthplace Memorial and Urban Design of the Surroundings
项目地点 / Location：河南洛阳 / Luoyang, Henan
设计时间 / Design Period：2019.8—
用地面积 / Site Area：10,677 m²
建筑面积 / Floor Area：10,311 m²
建筑设计 / Architecture：李兴钢、魏鸣宇、江昊懋 / Li Xinggang, Wei Mingyu, Jiang Haomao

国家植物博物馆（竞赛方案）
National Botanical Museum (Competition)
项目地点 / Location： 云南昆明 / Kunming, Yunnan
设计时间 / Design Period： 2019.9—2019.11
用地面积 / Site Area： 285,498 m²
建筑面积 / Floor Area： 60,300 m²
建筑设计 / Architecture： 李兴钢、张音玄、易灵洁、王汉、魏鸣宇、苏杭、陆婧瑶、攸然（云南省设计院集团有限公司）、翟星玥（云南省设计院集团有限公司）/ Li Xinggang, Zhang Yinxuan, Yi Lingjie, Wang Han, Wei Mingyu, Su Hang, Lu Jingyao, You Ran (Yunnan Design Institute Group Co., Ltd.), Zhai Xingyue (Yunnan Design Institute Group Co., Ltd.)
结构设计 / Structure： 董越 / Dong Yue

上海临港星空之境公园综合服务中心
Service Center in "Starland" of Lin-Gang Free Trade Zone in Shanghai
项目地点 / Location： 上海 / Shanghai
设计时间 / Design Period： 2019.11—
用地面积 / Site Area： 1,084m²
建筑面积 / Floor Area： 1,242m²
建筑设计 / Architecture： 李兴钢、孙知行、苏杭 / Li Xinggang, Sun Zhixing, Su Hang
结构设计 / Structure： 夏兵 / Xia Bing
机电设计 / Engineering： 刘权熠、徐红星、高志宏 / Liu Quanyi, Xu Hongxing, Gao Zhihong
照明顾问 / Lighting： 张昕照明设计工作室 / Zhang Xin Lighting Studio

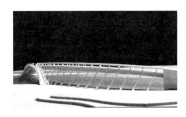

上海临港星空之境公园日月交辉桥
The Sun-Moon Splendour Bridge in "Starland" of Lin-Gang Free Trade Zone in Shanghai
项目地点 / Location： 上海 / Shanghai
设计时间 / Design Period： 2019.11—
建筑设计 / Architecture： 李兴钢、孙知行、滕凌霄 / Li Xinggang, Sun Zhixing, Teng Lingxiao
结构设计 / Structure： 汪明、谭晋鹏、夏兵 / Wang Ming, Tan Jinpeng, Xia Bing
机电设计 / Engineering： 刘权熠、张拗凡 / Liu Quanyi, Zhang Aofan
照明顾问 / Lighting： 张昕照明设计工作室 / Zhang Xin Lighting Studio

驻非盟使团新建馆舍（竞赛方案）
The New Embassy and Apartments for Mission of the People's Republic of China to the African Union (Competition)
项目地点 / Location： 埃塞俄比亚亚的斯亚贝巴 / Addis Ababa, Ethiopia
设计时间 / Design Period： 2020.4
用地面积 / Site Area： 40,000 m²
建筑面积 / Floor Area： 24,994 m²
建筑设计 / Architecture： 李兴钢、张音玄、邱涧冰、张哲、梁艺晓、李昀倩 / Li Xinggang, Zhang Yinxuan, Qiu Jianbing, Zhang Zhe, Liang Yixiao, Li Yunqian

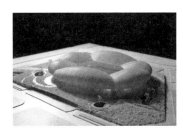

青岛亚洲杯足球场（竞赛方案）
Qingdao Football Stadium, AFC Asian Cup (Competition)
项目地点 / Location： 山东青岛 / Qingdao, Shandong
设计时间 / Design Period： 2020.4
用地面积 / Site Area： 174,277 m²
建筑面积 / Floor Area： 151,500 m²
建筑设计 / Architecture： 李兴钢、张音玄、魏鸣宇、腾凌霄、赵戈、陈曦、赵芹 / Li Xinggang, Zhang Yinxuan, Wei Mingyu, Teng Lingxiao, Zhao Ge, Chen Xi, Zhao Qin
结构设计 / Structure： 孙海林 / Sun Hailin

北京 2022 年冬奥会与冬残奥会延庆赛区场馆及设施——国家高山滑雪中心　◗

Beijing 2022 Olympic & Paralympic Winter Games Yanqing Zone Venues and Facilities — National Alpine Skiing Center

项目地点 / Location：北京 / Beijing

设计时间 / Design Period：2016.2—2018.12

施工时间 / Construction Period：2017.1—

用地面积 / Site Area：7,666,500 m²

建筑面积 / Floor Area：30,760 m²

总体规划 / Site Planning：李兴钢、盛况、邱涧冰、高治、洪于亮、王翔、崔志明、郝雯雯、刘晔、王萌、刘鹏、王陈栋 / Li Xinggang, Sheng Kuang, Qiu Jianbing, Gao Zhi, Hong Yuliang, Wang Xiang, Cui Zhiming, Hao Wenwen, Liu Ye, Wang Meng, Liu Peng, Wang Chendong

建筑设计 / Architecture：李兴钢、谭泽阳、梁旭、李欢、张捍平、沈周娅、张玹、宋檐 / Li Xinggang, Tan Zeyang, Liang Xu, Li Huan, Zhang Hanping, Shen Zhouya, Zhang Tao, Song Yan

结构设计 / Structure：任庆英、刘文珽、李森、杨松霖 / Ren Qingying, Liu Wenting, Li Sen, Yang Songlin

机电设计 / Engineering：申静、郝洁、霍新霖、祝秀娟、刘维、周蕾、张青、王旭、高学文 / Shen Jing, Hao Jie, Huo Xinlin, Zhu Xiujuan, Liu Wei, Zhou Lei, Zhang Qing, Wang Xu, Gao Xuewen

总图设计 / Master Plan：路建旗、高伟 / Lu Jianqi, Gao Wei

室内设计 / Interior：曹阳、马萌雪、闫宽 / Cao Yang, Ma Mengxue, Yan Kuan

景观设计 / Landscape：史丽秀、朱燕辉、关午军 / Shi Lixiu, Zhu Yanhui, Guan Wujun

照明顾问 / Lighting：丁志强 / Ding Zhiqiang

交通设计 / Traffic：洪于亮、叶平一、吴哲凌 / Hong Yuliang, Ye Pingyi, Wu Zheling

岩土工程 / Geotechnical：孙帅勤、李永东 / Sun Shuaiqin, Li Yongdong

北京 2022 年冬奥会与冬残奥会延庆赛区场馆及设施——国家雪车雪橇中心　◗

Beijing 2022 Olympic & Paralympic Winter Games Yanqing Zone Venues and Facilities — National Sliding Center

项目地点 / Location：北京 / Beijing

设计时间 / Design Period：2016.2—2018.12

施工时间 / Construction Period：2017.1—

用地面积 / Site Area：7,666,500 m²

建筑面积 / Floor Area：59,363 m²

总体规划 / Site Planning：李兴钢、盛况、邱涧冰、高治、洪于亮、王翔、崔志明、郝雯雯、刘晔、王萌、刘鹏、王陈栋 / Li Xinggang, Sheng Kuang, Qiu Jianbing, Gao Zhi, Hong Yuliang, Wang Xiang, Cui Zhiming, Hao Wenwen, Liu Ye, Wang Meng, Liu Peng, Wang Chendong

建筑设计 / Architecture：李兴钢、邱涧冰、张玉婷、刘紫骐、朱伶俐、刘扬、袁智敏、李碧舟、杨茹、王翔、姜汶林、陆婧瑶、陈译民 / Li Xinggang, Qiu Jianbing, Zhang Yuting, Liu Ziqi, Zhu Lingli, Liu Yang, Yuan Zhimin, Li Bizhou, Yang Ru, Wang Xiang, Jiang Wenlin, Lu Jingyao, Chen Yimin

结构设计 / Structure：任庆英、刘文珽、李正、张晓萌 / Ren Qingying, Liu Wenting, Li Zheng, Zhang Xiaomeng

机电设计 / Engineering：申静、李茂林、杨瀚宇、梁岩、祝秀娟、侯昱晟、张袆琦、张青、王旭、李宝华 / Shen Jing, Li Maolin, Yang Hanyu, Liang Yan, Zhu Xiujuan, Hou Yusheng, Zhang Yiqi, Zhang Qing, Wang Xu, Li Baohua

总图设计 / Master Plan：高治、朱庚鑫 / Gao Zhi, Zhu Gengxin

室内设计 / Interior：曹阳、张超 / Cao Yang, Zhang Chao

景观设计 / Landscape：史丽秀、关午军、朱燕辉 / Shi Lixiu, Guan Wujun, Zhu Yanhui

照明顾问 / Lighting：丁志强、李占杰 / Ding Zhiqiang, Li Zhanjie

交通设计 / Traffic：洪于亮、叶平一、吴哲凌 / Hong Yuliang, Ye Pingyi, Wu Zheling

岩土工程 / Geotechnical：孙帅勤 / Sun Shuaiqin

桥梁工程 / Bridge engineering：赵宏伟、迟啸起 / Zhao Hongwei, Chi Xiaoqi

北京 2022 年冬奥会与冬残奥会延庆赛区场馆及设施——配套设施　◗

Beijing 2022 Olympic & Paralympic Winter Games Yanqing Zone Venues and Facilities — Facilities

项目地点 / Location：北京 / Beijing

设计时间 / Design Period：2016.2—2018.12

施工时间 / Construction Period：2017.1—

用地面积 / Site Area：7,666,500 m²

建筑面积 / Floor Area：35,061 m²

建筑设计 / Architecture：李兴钢、邱涧冰、梁旭、刘振、张捍平、许乃天、万子昂、宋檐、苏杭 / Li Xinggang, Qiu Jianbing, Liang Xu, Liu Zhen, Zhang Hanping, Xu Naitian, Wan Zi'ang, Song Yan, Su Hang

结构设计 / Structure：任庆英、刘文珽、杨松霖、周轶伦 / Ren Qingying, Liu Wenting, Yang Songlin, Zhou Yilun

机电设计 / Engineering：郝洁、张超、霍新霖、祝秀娟、王志刚、侯昱晟、张青、王旭 / Hao Jie, Zhang Chao, Huo Xinlin, Zhu Xiujuan, Wang Zhigang, Hou Yusheng, Zhang Qing, Wang Xu

总图设计 / Master Plan：高治、郝雯雯、李爽、吴耀懿 / Gao Zhi, Hao Wenwen, Li Shuang, Wu Yaoyi

景观设计 / Landscape：史丽秀、朱燕辉、关午军 / Shi Lixiu, Zhu Yanhui, Guan Wujun

交通设计 / Traffic：洪于亮、叶平一、吴哲凌 / Hong Yuliang, Ye Pingyi, Wu Zheling

岩土工程 / Geotechnical：刘立健 / Liu Lijian

北京 2022 年冬奥会与冬残奥会延庆赛区场馆及设施——延庆冬奥村 ◗

Beijing 2022 Olympic & Paralympic Winter Games Yanqing Zone Venues and Facilities — Yanqing Olympic Village

项目地点 / Location：北京 / Beijing

设计时间 / Design Period：2016.2—2018.12

施工时间 / Construction Period：2017.1—

用地面积 / Site Area：34,3000 m²

建筑面积 / Floor Area：118,100 m²

总体规划 / Site Planning：李兴钢、盛况、邱涧冰、高治、洪于亮、王翔、崔志明、郝雯雯、刘晔、王萌、刘鹏、王陈栋 / Li Xinggang, Sheng Kuang, Qiu Jianbing, Gao Zhi, Hong Yuliang, Wang Xiang, Cui Zhiming, Hao Wenwen, Liu Ye, Wang Meng, Liu Peng, Wang Chendong

建筑设计 / Architecture：李兴钢、张音玄、张哲、张司腾、李虓、张一婷、梁艺晓、李慧、许乃天 / Li Xinggang, Zhang Yinxuan, Zhang Zhe, Zhang Siteng, Li Xiao, Zhang Yiting, Liang Yixiao, Li Hui, Xu Naitian

结构设计 / Structure：任庆英、刘文珽、王磊 / Ren Qingying, Liu Wenting, Wang Lei

机电设计 / Engineering：朱跃云、关若曦、张庆康、胡建丽、苏晓峰、全巍、曹磊、杨小雨、于征、高洁 / Zhu Yueyun, Guan Ruoxi, Zhang Qingkang, Hu Jiangli, Su Xiaofeng, Quan Wei, Cao Lei, Yang Xiaoyu, Yu Zheng, Gao Jie

总图设计 / Master Plan：刘晓琳、郝雯雯 / Liu Xiaolin, Hao Wenwen

室内设计 / Interior：曹阳、张洋洋、王强、张然 / Cao Yang, Zhang Yangyang, Wang Qiang, Zhang Ran

景观设计 / Landscape：史丽秀、关午军、王悦 / Shi Lixiu, Guan Wujun, Wang Yue

照明顾问 / Lighting：丁志强、黄星月 / Ding Zhiqiang, Huang Xingyue

交通设计 / Traffic：洪于亮、叶平一、吴哲凌 / Hong Yuliang, Ye Pingyi, Wu Zheling

岩土工程 / Geotechnical：刘立健 / Liu Lijian

北京 2022 年冬奥会与冬残奥会延庆赛区场馆及设施——延庆山地新闻中心 ◗

Beijing 2022 Olympic & Paralympic Winter Games Yanqing Zone Venues and Facilities — Yanqing Mountain Media Centre

项目地点 / Location：北京 / Beijing

设计时间 / Design Period：2016.2—2018.12

施工时间 / Construction Period：2017.1—

用地面积 / Site Area：34,3000 m²

建筑面积 / Floor Area：194,000 m²

总体规划 / Site Planning：李兴钢、盛况、邱涧冰、高治、洪于亮、王翔、崔志明、郝雯雯、刘晔、王萌、刘鹏、王陈栋 / Li Xinggang, Sheng Kuang, Qiu Jianbing, Gao Zhi, Hong Yuliang, Wang Xiang, Cui Zhiming, Hao Wenwen, Liu Ye, Wang Meng, Liu Peng, Wang Chendong

建筑设计 / Architecture：李兴钢、张音玄、闫昱、杨曦、王思莹、胡家源、田甜、张钊 / Li Xinggang, Zhang Yinxuan, Yan Yu, Yang Xi, Wang Siying, Hu Jiayuan, Tian Tian, Zhang Zhao

结构设计 / Structure：任庆英、刘文珽、张雄迪、李路彬 / Ren Qingying, Liu Wenting, Zhang Xiongdi, Li Lubin

机电设计 / Engineering：申静、李茂林、霍新霖、祝秀娟、周蕾、张青、王旭、王浩 / Shen Jing, Li Maolin, Huo Xinlin, Zhu Xiujuan, Zhou Lei, Zhang Qing, Wang Xu, Wang Hao

总图设计 / Master Plan：郝雯雯 / Hao Wenwen

室内设计 / Interior：曹阳、李毅 / Cao Yang, Li Yi

景观设计 / Landscape：关午军、张宛岚、常琳 / Guan Wujun, Zhang Wanlan, Chang Lin

照明顾问 / Lighting：丁志强、黄星月 / Ding Zhiqiang, Huang Xingyue

交通设计 / Traffic：洪于亮、叶平一、吴哲凌 / Hong Yuliang, Ye Pingyi, Wu Zheling

岩土工程 / Geotechnical：周超华 / Zhou Chaohua

出版与发表
PUBLICATIONS AND ARTICLES

出版 / Publications

李兴钢 . 静谧与喧嚣 [M]. 北京：中国建筑工业出版社，2015.6.

LI Xinggang. Tranquility and Noise[M]. Beijing: China Architecture & Building Press, 2015.6.

李兴钢 . 李兴钢 2004-2013 胜景几何 [J]. 城市环境设计 , 2014, No.079(01)

LI Xinggang. LI XINGGANG 2004-2013 Geometry and Sheng Jing[J]. URBAN ENVIRONMENT DESIGN, 2014, No.001(04)

李兴钢 . 李兴钢 [M]. 北京：中国建筑工业出版社，2012.6.

LI Xinggang. LI XINGGANG[M]. Beijing: China Architecture & Building Press, 2012.6.

李兴钢 . 织梦筑鸟巢 国家体育场——设计篇 [M]. 北京：中国建筑工业出版社，2009.12.

LI Xinggang. Building of the Dreamed Bird Nest: National Stadium - Chapter of Design[M]. Beijing: China Architecture & Building Press, 2009.12.

发表 / Articles

李兴钢，侯新觉 . 万峰林中的石头房子——贵州兴义楼纳建筑师公社露营服务中心 [J]. 时代建筑 , 2020, No.173(05): 82-91.

LI Xinggang, HOU Xinjue. A Stone House in Lofty Hills: The Camping Base Service Center in Louna, Xingyi, Guizhou Province[J]. TIME + ARCHITECTURE, 2020, No.173(05): 82-91.

李兴钢 . 与自然交互的建筑 [J]. 当代建筑 , 2020, No.001(1): 16-20.

LI Xinggang. Interaction with Nature Through Architecture[J]. CONTEMPORARY ARCHITECTURE, 2020, No.001(1): 16-20.

李兴钢，郑旭航 . "仓阁"：废弃工业建筑的新生——北京首钢工舍智选假日酒店设计 [J]. 建筑学报 , 2019, No.61310: 81-85.

LI Xinggang, ZHENG Xuhang. CangGe: The Rebirth of Abandoned Factories on the Design of Holiday Inn Express Beijing Shougang Silo-Pavilion[J]. ARCHITECTURAL JOURNAL, 2019, No.61310: 81-85.

张音玄，李兴钢，梁旭 . 走向公共的市政设施——北京市朝阳区生活垃圾综合处理厂焚烧中心 [J]. 建筑学报 , 2019, No.608(05): 68-71.

ZHANG Yinxuan, LI Xinggang, LIANG Xu. Infrastructure Accessible to the Public-on the Incineration Center of Domestic Waste Comprehensive Treatment Plant at Chaoyang District, Beijing[J]. ARCHITECTURAL JOURNAL, 2019, No.608(05): 68-71.

李兴钢 . 文化维度下的冬奥会场馆设计——以北京 2022 冬奥会延庆赛区为例 [J]. 建筑学报 , 2019, No.604(01): 35-42.

LI Xinggang. The Design of Stadiums for Winter Olympics in a Cultural Dimension: Taking the Yanqing Competition Zone of Beijing 2022 as an Example[J]. ARCHITECTURAL JOURNAL, 2019, No.604(01): 35-42.

李兴钢 . 佛光寺的启示——一种现实理想空间范式 [J]. 建筑学报 , 2018, No.600(09): 28-33.

LI Xinggang. Enlightment from the Foguangsi: A Paradigm for Realistic Ideal Space[J]. ARCHITECTURAL JOURNAL, 2018, No.600(09): 28-33.

李兴钢 . 身临其境，胜景几何 "微缩北京" / 大院胡同 28 号改造 [J]. 时代建筑 , 2018, No.162(04): 84-95.

LI Xinggang. Immersive in the Integrated Geometry and Poetic Scenery: "Miniature Beijing" at No.28 Dayuan Hutong[J]. Time + Architecture, 2018, No.162(04): 84-95.

李兴钢 . 胜景几何：人工与自然的互成 [J]. 建筑设计 ,2018, No.88(6):16-23.

LI Xinggang, Integrating Geometry Within a Poetic Setting: Steering a Path Between the Artificial and the Natural[J]. Architectural Design, 2018, No.88(6):16-23.

李兴钢，姜汶林 . "看我所见"——巴拉甘阅读报告 [J]. 建筑学报 , 2018, No.596(05): 83-90.

LI Xinggang, JIANG Wenlin. "Do not look at what I do, see what I see" - Research on Luis Barragan[J]. ARCHITECTURAL JOURNAL, 2018, No.596(05): 83-90.

李兴钢，侯新觉，谭舟 . "微缩北京"——大院胡同 28 号改造 [J]. 建筑学报 , 2018, No.598(07): 5-15.

LI Xinggang, HOU Xinjue, TAN Zhou. Miniature Beijing: The Renovation of No.28 Dayuan Hutong[J]. ARCHITECTURAL JOURNAL, 2018, No.598(07): 5-15.

范路，李兴钢 . 静谧胜景与诗意几何——建筑师李兴钢访谈 [J]. 建筑师 , 2018, No.192(02): 6-13.

FAN Lu, LI Xinggang. Tranquil Scene and Poetic Geometry: An Interview with Architect Li Xinggang[J]. The Architect, 2018, No.192(02): 6-13.

李兴钢 . 李兴钢自述 [J]. 世界建筑 , 2017, No.323(05): 65+141.

LI Xinggang. Li Xinggang[J], World Architect, 2017, No.323(05): 65+141.

李兴钢 . 作为"介质"的结构——天津大学新校区综合体育馆设计 [J]. 建筑学报 , 2016, No.579(12): 62-65.

LI Xinggang. Structure as Medium: The Gymnasium of New Campus of Tianjin University[J]. ARCHITECTURAL JOURNAL, 2016, No.579(12): 62-65.

李兴钢 . 绩溪博物馆 [J]. a+u, 2016, 546(2): 90-95.

LI Xinggang. Jixi Museum[J]. a+u, 2016, 546(2): 90-95.

李兴钢 , 李喆 . 微缩之城——商丘博物馆的设计与建造 [J]. 建筑学报 , 2016, No.570(03): 55-59.

LI Xinggang, LI Zhe. Miniature City - The Design and Construction of Shangqiu Museum[J]. ARCHITECTURAL JOURNAL, 2016, No.570(03): 55-59.

李兴钢 . 瞬时桃花源 , 南京 [J]. domus, 2015, 100(8): 80-89.

LI Xinggang. INSTANT PEACH BLOSSOM SPRING, NANJING[J]. domus, 2015, 100(8): 80-89.

李兴钢 , 张玉婷 , 姜汶林 . 瞬时桃花源 [J]. 建筑学报 , 2015, No.566(11): 30-39.

LI Xinggang, ZHANG Yuting, JIANG Wenlin. The Instant Peach Blossom Spring Nanjing, Jiangsu[J]. ARCHITECTURAL JOURNAL, 2015, No.566(11): 30-39.

李兴钢 . 静谧与喧嚣 [J]. 建筑学报 , 2015, No.566(11): 40-43.

LI Xinggang. Tranquility and Noise[J]. ARCHITECTURAL JOURNAL, 2015, No.566(11): 40-43.

李兴钢 . Archi-neering: 营造几例 [J]. 建筑师 , 2015, No.174(02): 122-125.

LI Xinggang. Archi-neering: Samples of Ying Zao[J]. The Architect, 2015, No.174(02): 122-125.

李兴钢 , 张音玄 , 张哲 , 邢迪 . 留树作庭随遇而安折顶拟山会心不远——记绩溪博物馆 [J]. 建筑学报 , 2014, No.546(02): 40-45.

LI Xinggang, ZHANG Yinxuan, ZHANG Zhe, XING Di. Tree, Garden, and Mountain-Like Roofs on Jixi Museum[J]. ARCHITECTURAL JOURNAL, 2014, No.546(02): 40-45.

黄居正 , 易娜 , 赵辰 , 鲁安东 , 董豫赣 , 庄慎 , 金秋野 , 黄涛英 , 李兴钢 , 任浩 , 张音玄 , 邢迪 . 瓦壁当山——李兴钢绩溪博物馆研讨会纪要 [J]. 建筑师 , 2014, No.167(01): 98-125.

HUANG Juzheng, YI Na, ZHAO Chen, LU Andong, DONG Yugan, ZHUANG Shen, JIN Qiuye, HUANG Taoying, LI Xinggang, REN Hao, ZHANG Yinxuan, XING Di. Tiled Wall as Mountain: The Seminar of Li Xinggang's Jixi Museum[J]. The Architect, 2014, No.167(01): 98-125.

李兴钢 , 易灵洁 . 建筑实践中的材料选择 [J]. 时代建筑 , 2014, No.137(03): 50-57.

LI Xinggang, YI Lingjie. Selection of Materials in Architectural Practice[J]. TIME +ARCHITECTURE, 2014, No.137(03): 50-57.

赵迪 , 李兴钢 , 周榕 . 遂心 , 随性——周榕对话李兴钢 [J]. 城市环境设计 , 2014, No.083(06): 216-219.

ZHAO Di, LI Xinggang, ZHOU Rong. Follow the Heart, Along with Nature-Dialogue Between Zhao Rong And Li Xinggang[J]. URBAN ENVIRONMENT DESIGN, 2014, No.083(06): 216-219.

李兴钢 , 谭泽阳 , 邱涧冰 . 国家体育场的赛后利用 [J]. 世界建筑 , 2013, (08): 38-51+128.

LI Xinggang, TAN Zeyang, QIU Jianbing. The Post-Olympics Usage of the National Stadium[J]. World Architect, 2013, (08): 38-51+128.

李兴钢 . 胜景几何 [J]. 建筑技艺 , 2013, No.218(05): 26-28.

LI Xinggang. SHENG JING JI HE[J]. Architecture Technique, 2013, No.218(05): 26-28.

李兴钢 , 易灵洁 . 大草原上的"小帐篷"——图记元上都遗址工作站 [J]. 建筑学报 , 2013, No.533(01): 52-59.

LI Xinggang, YI Lingjie. "Small Tent" on Grasslands: Notes of the Image Design for the Entrance for Site of Xanadu[J]. ARCHITECTURAL JOURNAL, 2013, No.533(01): 52-59.

李兴钢 , 马津 . 新小镇 , 新希望——西柏坡华润希望小镇（一期）设计感悟 [J]. 城市建筑 , 2013, No.105(01): 94-101.

LI Xinggang, MA Jin. New Town, New Hope: China Resources Hope Town, Xibaipo (Phase 1)[J]. URBANISM AND ARCHITECTURE, 2013, No.105(01): 94-101.

李兴钢 . 第一见证："鸟巢"的诞生、理念、技术和时代决定性 [D]. 天津大学 ,2012.

LI Xinggang. First Witness: Birth, Concept, Technology and Time Decisiveness of "Bird Nest"[D]. Tianjin University, 2012.

李兴钢 , 谭泽阳 , 张玉婷 . 探求建筑形式、结构与空间的同一性——海南国际会展中心设计手记 [J]. 建筑学报 , 2012, No.527(07): 44-47.

LI Xinggang, TAN Zeyang, ZHANG Yuting. Explore the Unity of Architecture Form, Structure, and Space: Notes on the Design of Hainan International Conference & Exhibition Center[J]. ARCHITECTURAL JOURNAL, 2012, No.527(07): 44-47.

李兴钢 , 付邦保 , 张音玄 , 谭泽阳 . 虚像、现实与灾难体验——建川镜鉴博物馆暨汶川地震纪念馆设计 [J]. 建筑学报 , 2010, No.507(11): 44-47.

LI Xinggang, FU Bangbao, ZHANG Yinxuan, TAN Zeyang. Virtual Image, Reality and Disaster Experience: Design of Jianchuan Mirror Museum & Wenchuan Earthquake Memorial[J]. ARCHITECTURAL JOURNAL, 2010, No.507(11): 44-47.

赵夏榕 . 从本原出发——访中国建筑设计研究院副总建筑师、李兴钢建筑设计工作室主持人 李兴钢 [J]. 设计家 , 2010, No.45(02): 34-41.

ZHAO Xiarong. Starting Point & Endpoint: the Essence of Human Activities and Culture[J]. DESIGNER & DESIGNING, 2010, No.45(02): 34-41.

李兴钢 . 我看膜结构 [J]. 世界建筑 , 2009, No.232(10): 30-31.

LI Xinggang. My View on Membrane Architecture[J]. World Architect, 2009, No.232(10): 30-31.

李兴钢 . 由国家体育场的设计看建筑向本原回归的倾向 [J]. 世界建筑 , 2008, No.216(06): 22-31.

LI Xinggang. The Tendency to Trace the Essence of Architecture from the Design of the National Stadium[J]. World Architect, 2008, No.216(06): 22-31.

李兴钢 . 喧嚣与静谧 [J]. 建筑师 , 2008, No.136(06): 110-114.

LI Xinggang. Noise and Tranquility[J]. The Architect, 2008, No.136(06): 110-114.

李兴钢 , 张音玄 , 付邦保 . 表皮与空间——北京复兴路乙 59-1 号改造 [J]. 建筑学报 , 2008, No.484(12): 58-64.

LI Xinggang, ZHANG Yinxuan, FU Bangbao. Facade and Space: Renovation of B59-1 Fuxing Road, Beijing[J]. ARCHITECTURAL JOURNAL, 2008, No.484(12): 58-64.

吴洪德 . 自返其身的建筑工作 国家体育场 "鸟巢" 中方总建筑师李兴钢访谈 [J]. 时代建筑 , 2008, No.102(04): 42-51.

WU Hongde. Self-reflexive Architecture Work: Interviewing with Li Xinggang on the National Stadium of China, the "Bird's Nest"[J]. TIME +ARCHITECTURE, 2008, No.102(04): 42-51.

李兴钢 . 国家体育场设计 [J]. 建筑学报 , 2008, No.480(08): 1-17.

LI Xinggang. Design of the National Stadium of China[J]. ARCHITECTURAL JOURNAL, 2008, No.480(08): 1-17.

李兴钢 , 苗苗 . 北京西直门交通枢纽设计研究 [J]. 世界建筑 , 2008, No.218(08): 50-61.

LI Xinggang, MIAO Zhuo. The Design & Research of the Beijing Xizhimen Traffic Hub[J]. World Architect, 2008, No.218(08): 50-61.

李兴钢 . 威尼斯的纸砖房 [J]. 城市建筑 , 2008,No.51(12): 42-45.

LI Xinggang. Paper-brick House in Venice[J]. URBANISM AND ARCHITECTURE, 2008,No.51(12): 42-45.

李兴钢 . 新理念、新材料、新技术、新方法在国家体育场设计中的运用 [J]. 建筑创作 , 2007, No.97(07). 68-83.

LI Xinggang. The Application of New Idea, Materials, Technology and Method in the Design of the State Stadium[J]. ARCHICREATION. 2007, No.97(07): 68-83.

李兴钢 . 北京大兴区文图馆 [J]. 世界建筑 , 2006, (10): 128-133.

LI Xinggang. Beijing Daxing District Cultural Library[J]. World Architect, 2006, (10): 128-133.

王胜霞 , 刘靖怡 . 发现建筑——李兴钢自述 [J]. 城市环境设计 , 2004, (02): 109-116.

WANG Shengxia, LIU Jingyi. Discover Architecture[J]. URBAN ENVIRONMENT DESIGN, 2004, (02): 109-116.

李兴钢 . "城市化" 的建筑——北京兴涛会馆 [J]. 世界建筑 , 2002, (01): 87-88.

LI Xinggang. "Urbanized" Building: Beijing Xingtao Club[J]. World Architect, 2002, (01): 87-88.

范雪 . 两个作品 三个话题——李兴钢访谈 [J]. 建筑学报 , 2002, (08): 41-42.

FAN Xue. Two Projects and Three Topics: Interview with Li Xinggang[J]. ARCHITECTURAL JOURNAL, 2002, (08): 41-42.

李兴钢 . 追寻本原的学校 [J]. 建筑学报 , 2001, (12): 14-17.

LI Xinggang. A School Seeking the Origin[J]. ARCHITECTURAL JOURNAL, 2001, (12): 14-17.

李兴钢 , 马先 . 城镇型高级住宅的设计研究和实践 [J]. 小城镇建设 , 2001, (07): 37-39.

LI Xinggang, MA Xian. Research and Practice on the Design of High-Grade Residences Oriented to Cities and Towns[J]. DEVELOPMENT OF SMALL CITIES & TOWNS, 2001, (07): 37-39.

李兴钢 , 李靖 . 运动的墙板·运动的人——北京兴涛展示接待中心 [J]. 建筑与设计 , 2001, (04): 18-21.

LI Xinggang, Li Jing. Moving Walls and Moving People: Beijing Xingtao Exhibiton & Reception Center[J]. ARCHITECTURE & DESIGN, 2001, (04): 18-21.

李兴钢 . 小区规划的结构设计方法探讨 [J]. 建筑师 , 2000, 93(4): 16-35.

LI Xinggang. Explore the Structure Design Methods of Residential Community Planning[J]. ARCHITECT, 2000, 93(4): 16-35.

李兴钢 . "城市" 与 "建筑" ——兼介北京兴涛学校 [J]. 建筑学报 , 1999, (12): 33-34.

LI Xinggang. "City" and "Building": Introduction to Beijing Xingtao School[J]. ARCHITECTURAL JOURNAL, 1999, (12): 33-34.

王群 , 李兴钢 . 北京金阳大厦设计 . 建筑师 , 1998, 83(8): 23-25.

WANG Qun, LI Xinggang. The Design of Beijing Jinyang Building[J]. ARCHITECT, 1998, 83(8): 23-25.

李兴钢 . "场所" 中的建筑·关于 "原型" 的思考 [J]. 建筑学报 , 1997, (07): 38-41.

LI Xinggang. Buildings on "Site": Thoughts on the "Prototype" [J]. ARCHITECTURAL JOURNAL, 1997, (07): 38-41.

李兴钢 . 北京兴涛居住小区规划 [J]. 建筑学报 , 1996, (10): 29-31.

LI Xinggang. Planning of Beijing Xingtao Residential Community[J]. ARCHITECTURAL JOURNAL, 1996, (10): 29-31.

展览
EXHIBITIONS

2019 长江美术馆开馆展，山西太原
Yangtze Art Museum Opening Exhibition, Taiyuan, Shanxi
"新生于旧"，2019 北京国际设计周城市更新主题展，北京
"Urban Renewal", Theme Exhibition of 2019 Beijing Design Week, Beijing
"柒作——东方新营造"，2019 北京国际设计周城市更新主题展，北京
"Seven Works of China New Architecture", Theme Exhibition of 2019 Beijing Design Week, Beijing

2017 第二十六届世界建筑大会"融—合之间"中国建筑展，韩国首尔
Between the Fusion, China Architecture Exhibition, Seoul, South Korea
"当代中国建筑"克拉科夫建筑双年展，波兰克拉科夫
Contemporary Chinese Architecture, Krakow Architecture Biennale 2017, Krakow, Poland
库里奇巴国际艺术双年展"地平线——中国当代建筑展"，巴西库里奇巴
Horizon — Chinese Contemporary Architectural Exhibition，Brazil, Curitiba
第九届威海国际人居节"剖面——一本杂志视角下的中国建筑设计五年"，山东威海
The 9th Weihai International Habitat Festival，"Section — five years of China Architecture", Weihai, Shandong
"大国工匠 - 校友成果"展（天津大学建筑学院 80 周年院庆活动），天津
Great Country, Great Craftsman: Alumnus Works Exhibition of Tianjin University, Tianjin
深圳—香港城市 \ 建筑双城双年展，广东深圳 / 香港
SZ—HK Urbanism\ Architecture Bi-City Biennale, Shenzhen, Guangdong / Hong Kong

2016 哈佛大学 GSD "中国当代建筑展"，美国马萨诸塞州
Harvard GSD Chinese Contemporary Architecture Exhibition, Massachusetts, USA
中澳设计研究对话展，澳大利亚墨尔本
China-Australian Design Research Dialogue, Melbourne, Australia
"格物"展，上海
"Investigate it", Shanghai
"叠合院"——护国寺西巷 37 号院更新改造设计展暨智库咖啡展，北京
"The Stack Courtyard House" — Renovation of West Huguosi St. 37, Beijing

2015 建筑中国 1000 展览，北京
Architecture China 1000, Beijing
存续：中国建筑实践调研展，北京
Duration: An Investigation on Architecture Practice in China, Beijing
"结构建筑学"展，北京
Archi-Neering Design Exhibition, Beijing
10 × 100——UED 十年百名建筑师展，北京
10 × 100, UED Exhibition, 10 Years, 100 Architects, Beijing

2014 "造 / 建筑中国"——中法建交 50 周年纪念展，法国里昂
ZAO / Architecture China, La c é l é bration du 50 è me anniversaire des relations diplomatiques entre la France et la Chine, Lyon, France

2013 中国宫——建筑中国 2013，西班牙塞戈维亚
Palace of China — Architecture China 2013, Segovia, Spain
"胜景几何"微展，北京
"Sheng Jing Ji He", Beijing/China

2013 西岸 2013：建筑与当代艺术双年展，上海
West Bund 2013: A Biennial of Architecture and Contemporary Art, Shanghai
2012 从北京到伦敦——当代中国建筑展，英国伦敦
From Beijing to London: Contemporary Chinese Architecture, London, England
建筑中国 100 展，德国曼哈姆
Architecture China: The 100 Contemporary Projects, Manheim, Germany
2011 "向东方——中国建筑景观"展，意大利罗马
Verso Est, New Chinese Architectural Landscape 2011, Rome, Italy
2010 "后实验时代的中国地域建筑"展，卡尔斯鲁厄 / 布拉格
Chinese Regional Architecture in a Post-Experimental Age, Karlsruhe / Prague
"东风——中国新建筑 2000—2010"展，瑞士巴塞尔 / 列支敦士登瓦杜兹
Dongfeng, New China Construction 2000—2010, Basel, Switzerland / Vaduz, Liechtenstein
心造——中国当代建筑前沿展，比利时布鲁塞尔
"Heart-Made", The Cutting-Edge of Chinese Contemporary Architecture, Brussels, Belgium
首都第十七届规划设计方案汇报展，北京
The 17th Capital Planning and Design Proposal, Beijing
2008 第 11 届威尼斯国际建筑双年展，意大利威尼斯
The 11th Venice Biennale of Architecture, Venice, Italy
"从幻象到现实：活的中国园林"展，德国德累斯顿
"Illusion Into Reality: Chinese Gardens for Living", Dresden, Germany
2007 深圳—香港城市 \ 建筑双城双年展，广东深圳 / 香港
SZ—HK Urbanism \ Architecture Bi-City Biennale, Shenzhen, Guangdong / Hong Kong
北京大声艺术展，北京
"Get It Louder", Beijing
"发生"——北京左右艺术区艺术展，北京
"Happen", Left & Right Art Zone, Beijing
2005 "状态"——中国当代青年建筑师作品八人展，北京
"Status", Eight Young Chinese Architects, Beijing
首届深圳城市 \ 建筑双年展，广东深圳
The 1st Shenzhen Urbanism \ Architecture Biennale, Shenzhen, Guangdong

获奖
AWARDS

项目获奖 / Project Awards

2019   "元上都遗址博物馆" 获亚洲建筑师协会建筑奖金奖

Gold of ARCASIA Awards for Architecture, Museum for Site of XANADU

"绩溪博物馆" 获中国建筑学会建筑创作大奖（2009-2019）

Architectural Creation Award of the Architectural Society of China (2009-2019), Jixi Museum

"海南国际会展中心" 获中国建筑学会建筑创作大奖（2009-2019）

Architectural Creation Award of the Architectural Society of China (2009-2019), Hainan International Convention & Exhibition Center

"天津大学新校区综合体育馆" 获中国建筑学会建筑创作大奖（2009-2019）

Architectural Creation Award of the Architectural Society of China (2009-2019), Gymnasium of the New Campus of Tianjin University

"唐山'第三空间'" 获中国建筑学会建筑创作大奖（2009-2019）

Architectural Creation Award of the Architectural Society of China (2009-2019), The "Third Space" in Tangshan

"元上都遗址博物馆" 获中国建筑学会建筑创作大奖（2009-2019）

Architectural Creation Award of the Architectural Society of China (2009-2019), Museum for Site of XANADU

2018   "天津大学新校区综合体育馆" 获 ArchDaily2018 年度全球建筑大奖

Global Architecture Award for Sports Buildings of 2018 ArchDaily Building of the Year Award, Gymnasium of the New Campus of Tianjin University

"微缩北京"——大院胡同 28 号院改造获 "WA 中国建筑奖" 居住贡献奖优胜奖

Honorable Prize of Housing Award of 2018 WA China Architecture Award, "Miniature Beijing", Renovation of No. 28 Dayuan Hutong

"微缩北京"——大院胡同 28 号院改造获 "WA 中国建筑奖" 设计实验奖入围奖

Finalist of Design Experiment Award of 2018 WA China Architecture Award, "Miniature Beijing", Renovation of No. 28 Dayuan Hutong

"天津大学新校区综合体育馆" 获中国建筑学会 2017-2018 年度建筑设计奖 建筑创作·公共建筑类金奖

Global Award of China Architecture Design Award of the Architectural Society of China(2017-2018) (Architectural Creation), Gymnasium of the New Campus of Tianjin University

"商丘博物馆" 获中国建筑学会 2017-2018 年度建筑设计奖 建筑创作银奖

Silver Award of China Architecture Design Award of the Architectural Society of China (2017-2018) (Architectural Creation), Shangqiu Museum

"唐山'第三空间'" 获中国建筑学会 2017-2018 年度建筑设计奖 居住建筑类银奖

Silver Award of China Architecture Design Award of the Architectural Society of China (2017-2018) (Residential Building Category), The "Third Space" in Tangshan

"商丘博物馆" 获中国建筑学会 2017-2018 年度建筑设计奖 建筑幕墙专业类一等奖

First Prize of China Architecture Design Award of the Architectural Society of China(2017-2018) (Building Facade Category) , Shangqiu Museum

"元上都遗址博物馆" 获中国建筑学会 2017-2018 年度建筑设计奖 建筑创作·公共建筑类银奖

Silver Award of China Architecture Design Award of the Architectural Society of China Architectural Creation in Public Building Category（2017-2018), Museum for Site of XANADU

2017   "建川镜鉴博物馆暨汶川地震纪念馆" 获第九届远东建筑奖提名奖

Finalist of the 9th Far East Architecture Award, Jianchuan Mirror Museum & Wenchuan Earthquake Memorial

"天津大学新校区综合体育馆" 获全国优秀工程勘察设计行业奖建筑工程公建类一等奖

First Prize of the National Excellent Engineering Investigation and Design Industry Award (Public Buildings of Architectural), Gymnasium of the New Campus of Tianjin University

"商丘博物馆" 获 2017 年北京市优秀工程勘察设计奖 综合奖（公共建筑）一等奖

First Prize of the Excellent Engineering Investigation and Design Award of Beijing, (Comprehensive Award (Public Buildings), Shangqiu Museum

2017 "天津大学新校区综合体育馆"获 2017 年北京市优秀工程勘察设计奖综合奖一等奖

First Prize of the Excellent Engineering Investigation and Design Award of Beijing (Comprehensive Award), Gymnasium of the New Campus of Tianjin University

"天津大学新校区综合体育馆"获"第四届中国建筑传媒奖"入围奖

Finalist of 2017 China Architecture Media Award, Gymnasium of the New Campus of Tianjin University

2016 "天津大学新校区综合体育馆"获"WA 中国建筑奖" 技术进步奖优胜奖

Honorable Prize of Technological Innovation Award of 2016 WA China Architecture Award, Gymnasium of the New Campus of Tianjin University

"唐山'第三空间'"获"WA 中国建筑奖"居住贡献奖优胜奖

Honorable Prize of Housing Award of 2016 WA China Architecture Award, The "Third Space" in Tangshan

"元上都遗址工作站"获"WA 中国建筑奖"建筑成就奖入围奖

Finalist of Architectural Achievement Award of 2016 WA China Architecture Award, Workstation for Site of XANADU

"天津大学新校区综合体育馆"获"自然建造——第四届中国建筑传媒奖技术探索奖"入围奖

Finalist of Technological Exploration Award of the 4th China Architecture Media Award, China Architecture Award, Gymnasium of the New Campus of Tianjin University

"绩溪博物馆"获中国建筑学会建筑创作奖金奖（公共建筑类）

Gold Award of Architectural Creation Award of the Architectural Society of China (Public Building Category), Jixi Museum

"鸟巢文化中心"获 2016 年度中国建筑学会建筑创作奖银奖（公共建筑类）

Silver Award of 2016 ASC Architectural Creation Award (Public Buildings), Birds' Nest Cultural Center

"绩溪博物馆"获中国建筑学会中国建筑设计奖

China Architecture Design Award (Architectural Creation) of 2016 China Architecture Design Award of the Architectural Society of China, Jixi Museum

2015 "绩溪博物馆"获全国优秀工程勘察设计行业奖（建筑工程）公建一等奖

First Prize of National Excellent Engineering Investigation and Design Industry Award (Architectural Engineering), Jixi Museum

"西柏坡华润希望小镇"获住房城乡建设部第一批田园建筑优秀作品一等奖

First Prize of The Excellent Garden Building Works of the Ministry of Housing and Urban-Rural Development, Xibaipo China Resources Hope Town

"绩溪博物馆"获北京市第十八届优秀工程设计一等奖

First Prize of the 18th Excellent Engineering Investigation and Design Award of Beijing, Jixi Museum

2014 "国家体育场（2008 年奥运会主体育场）"获"中国当代十大建筑"奖

Top Ten Modern Buildings of China, National Stadium

"绩溪博物馆"获"WA 中国建筑奖"城市贡献奖佳作奖

Honorable Prize of 2014 WA China Architecture Award, Jixi Museum

2013 "海南国际会展中心"获全国优秀工程勘察设计行业奖建筑工程公建一等奖

First Prize of the National Excellent Engineering Investigation and Design Industry Award (Public Buildings of Architectural Engineering), Hainan International Convention & Exhibition Center

"海南国际会展中心"获中国建筑设计奖（建筑创作）金奖

Gold Award for Architectural Creation of Award for Chinese Architecture, Hainan International Convention & Exhibition Center

"海南国际会展中心"获北京市第十七届优秀工程设计一等奖

First Prize of the 17th Excellent Engineering Investigation and Design Award of Beijing, Hainan International Convention & Exhibition Center

2012 "海南国际会展中心"获"WA 中国建筑奖"入围奖

Finalist of 2012 WA China Architecture Award, Hainan International Convention & Exhibition Center

"西柏坡华润希望小镇"获中国建筑传媒奖第三届居住建筑特别奖入围奖

Finalist of Special Award for The 3rd Residential Buildings of China Architecture Media Award, Xibaipo China Resources Hope Town

"昌平线一期工程"获北京市第十六届优秀工程设计一等奖

First Prize of the 16th Excellent Engineering Investigation and Design Award of Beijing, Changping Line of Beijing Subway (First-Stage)

"北京地铁昌平线朱辛庄站"获北京市第十六届优秀工程设计轨道交通站点设计单项奖

Individual Award of the 16th Excellent Engineering Investigation and Design Award of Beijing (Individual Award in Rail Transit Station Design or Rail Transit Station Design Individual Award), Zhuxinzhuang Station of Changping Line of Beijing Subway

2011 "北京复兴路乙 59—1 号改造"获第十四届全国优秀工程勘察设计奖银奖

Silver Award of The 14th National Excellent Engineering Investigation and Design Award, Renovation of No.B-59-1, Fuxing Road in Beijing

2011 　"北京地铁昌平线西二旗站"获国际建筑成就奖

International Achievement Awards, Xi'erqi Station of Changping Line of Beijing Subway

"北京地铁昌平线西二旗站"获新城轨道交通优秀设计评选"车站优秀设计"一等奖

First Prize in "Excellent Design of Station" of Appraisal and selection of Excellent Design of Rail Traffic in New Town of Beijing, Xi'erqi Station of Changping Line of Beijing Subway

"建川镜鉴博物馆暨汶川地震纪念馆"获第六届中国建筑学会建筑创作奖佳作奖

Architectural Creation Honorable Prize of the 6th ASC Architecture Award, Jianchuan Mirror Museum & Wenchuan Earthquake Memorial

"北京地铁4号线及大兴线地面出入口及附属设施"获北京市第十五届优秀工程设计一等奖

First Prize of the 15th Excellent Engineering Investigation and Design Award of Beijing, Accesses and Facilities of Line 4 & Daxing Line of Beijing Subway

"建川镜鉴博物馆暨汶川地震纪念馆"获北京市第十五届优秀工程设计评选一等奖

First Prize of the 15th Excellent Engineering Investigation and Design Award of Beijing, Jianchuan Mirror Museum & Wenchuan Earthquake Memorial

"威海"'Hiland·名座'"获第六届中国建筑学会建筑创作奖佳作奖

Architectural Creation Honorable Prize of the 6th ASC Architecture Award, "Hiland · Mingzuo" in Weihai

2010 　"建川镜鉴博物馆暨汶川地震纪念馆"获 THE CHICAGO ATHENAEUM 国际建筑奖

Chicago Athenaeum International Architecture Award, Jianchuan Mirror Museum & Wenchuan Earthquake Memorial

2009 　"国家体育场（2008年奥运会主体育场）"获国际奥委会和国际体育与休闲建筑协会金奖

Gold Award of IOC (International Olympic Committee) and IAKS (International Association for Sports and Leisure Facilities) Award, National Stadium

"国家体育场（2008年奥运会主体育场）"获国际奥委会和国际体育与休闲建筑协会残疾人联合会金奖

Gold Award of IPC (International Paralympic Committee) and IAKS (International Association for Sports and Leisure Facilities) Award, National Stadium

"国家体育场（2008年奥运会主体育场）"获英国皇家建筑师协会"莱伯金"建筑大奖

Architecture Award of Royal Institute of British Architects "RIBA" Architecture Award, National Stadium

"北京复兴路乙59-1号改造"获 THE CHICAGO ATHENAEUM 国际建筑奖

Chicago Athenaeum International Architecture Award, Renovation of No.B-59-1, Fuxing Road in Beijing

"威尼斯纸砖房"获 THE CHICAGO ATHENAEUM 国际建筑奖

Chicago Athenaeum International Architecture Award, Paper-brick House in Venice

"国家体育场（2008年奥运会主体育场）"获北京市奥运工程落实三大理念优秀勘察设计奖

Excellent Investigation and Design Award of Excellent Investigation and Design Award of Beijing for Implementing Three Concepts of Olympic Project, National Stadium

"北京复兴路乙59-1号改造"获2009年全国优秀工程勘察设计行业奖 建筑工程一等奖

First Prize in Architectural Engineering of 2009 National Excellent Engineering Investigation and Design Industry Award, Renovation of No.B-59-1, Fuxing Road in Beijing

"国家体育场（2008年奥运会主体育场）"获北京市第十四届优秀工程设计一等奖

First Prize of the 14th Excellent Engineering Investigation and Design Award of Beijing, National Stadium

"北京兴涛学校"获中国建筑学会建筑创作大奖

Creation Award of 2009 ASC Architectural Creation Award, Beijing Xingtao School

2008 　"国家体育场（2008年奥运会主体育场）"获2008年度全国优秀工程勘察设计奖金奖

Gold Award of 2008 National Excellent Engineering Investigation and Design Award, National Stadium

"北京复兴路乙59-1号改造"获第五届中国建筑学会建筑创作佳作奖

Honorable Prize of the 5th ASC Architectural Creation Award, Renovation of No.B-59-1, Fuxing Road in Beijing

"北京大兴区文化中心"获第五届中国建筑学会建筑创作佳作奖

Honorable Prize of the 5th ASC Architectural Creation Award, Daxing Cultural Center

"国家体育场（2008年奥运会主体育场）"获中国建筑学会建筑创作大奖（新中国成立60周年）

2008 Architectural Creation Award of ASC Architectural Creation Award（60th anniversary）, National Stadium

2006 　"天津泰达小学"获第四届中国建筑学会建筑创作奖佳作奖

Honorable Prize of the 4th ASC Architectural Creation Award, Teda Primary School

2003　"兴涛接待展示中心"入选 2003 年亚洲建筑师协会第七届建筑金奖选送项目
Selected project of Gold Award of the 7th ARCASIA Architecture, Xingtao Reception & Exhibition Center

2002　"兴涛接待展示中心"获英国"世界建筑奖"提名奖
Finalist of British "World Architecture Award", Xingtao Reception & Exhibition Center

"北京兴涛社区（一期）"获北京市第十届优秀工程设计一等奖
First Prize of the 10th Excellent Engineering Investigation and Design Award of Beijing, Beijing Xingtao Community (First-Stage)

2000　"北京兴涛学校"获第九届全国优秀工程勘察设计银奖
Silver Award of the 9th National Excellent Engineering Investigation and Design Award, Beijing Xingtao School

"北京兴涛学校"获建设部和非公交系统 2000 年优秀设计一等奖
First Prize of Excellent Design Award of 2000 Ministry of Construction and Non-Public Traffic System, Beijing Xingtao School

个人荣誉 / Individual Honour

2019　"北京科技盛典人物"称号
"Beijing Science and Technology Festival Figures"

2018　"最美科技工作者"称号
"Outstanding Science and Technology Workers"

2016　"全国工程勘察设计大师"称号
"National Engineering Survey and Design Master"

2012　当代中国百名建筑师
100 Architects of Contemporary China

2008　北京奥运工程建设功臣
Meritorious Architect of the Construction of Beijing Olympic

北京奥运会残奥会先进个人
Advanced Individual of Beijing Olympic and Paralympic

"科学中国人"称号
"Scientific Chinese"

2007　第十届中国青年科技奖
The 10th China Youth Science and Technology Award

2007 全球华人青年建筑师奖
2007 Global Young Chinese Architects Award

2006　新世纪百千万人才国家级人选
National Candidates for Hundreds of Millions of Talents in the New Century

全国五一劳动奖章
National Labor Medal

首都劳动奖章
Capital Labor Medal

第 16 届"中国十大杰出青年"提名奖
Award Nomination for The 16th "Ten Outstanding Young Chinese"

2005　国务院政府特殊津贴专家
Special Government Allowances

2004　第五届中国建筑学会青年建筑师奖
The 5th Young Architects Award by Architectural Society of China

亚洲建筑推动奖
Asian Architecture Development Award

工作室人员名单
COLLABORATORS

在职成员 / Working Stuff

李兴钢、谭泽阳、张音玄、邱涧冰、张哲、张玉婷、易灵洁、梁旭、邓建祥、张司腾、姜汶林、刘振、朱伶俐、孔祥惠、李欢、王汉、刘紫骐、侯新觉、张捍平、沈周娅、梁艺晓、袁智敏、谭舟、陆婧瑶、孙知行、陈译民、郭文嘉、魏鸣宇、苏杭、赵戈、滕凌霄、李昀倩

Li Xinggang, Tan Zeyang, Zhang Yinxuan, Qiu Jianbing, Zhang Zhe, Zhang Yuting, Yi Lingjie, Liang Xu, Deng Jianxiang, Zhang Siteng, Jiang Wenlin, Liu Zhen, Zhu Lingli, Kong Xianghui, Li Huan, Wang Han, Liu Ziqi, Hou Xinjue, Zhang Hanping, Shen Zhouya, Liang Yixiao, Yuan Zhimin, Tan Zhou, Lu Jingyao, Sun Zhixing, Chen Yimin, Guo Wenjia, Wei Mingyu, Su Hang, Zhao Ge, Teng Lingxiao, Li Yunqian

过往成员 / Former Employees

刘爱华、付邦保、李力、钟鹏、董烜、肖育智、王子耕、郭佳、薛从清、赵小雨、李宁、唐勇、孙鹏、董秀芳、邢迪、李喆、闫昱、陆少波、李碧舟、李慧

Liu Aihua, Fu Bangbao, Li Li, Zhong Peng, Dong Xuan, Xiao Yuzhi, Wang Zigeng, Guo Jia, Xue Congqing, Zhao Xiaoyu, Li Ning, Tang Yong, Sun Peng, Dong Xiufang, Xing Di, Li Zhe, Yan Yu, Lu Shaobo, Li Bizhou, Li Hui

研究生 / Graduate Students

戴泽均、张哲、弓蒙、朱磊、吴燕雯（联合培养）、张一婷、马津、尹璐（联合培养）、钟曼琳、王瑶（联合培养）、孟宁（联合培养）、李欢、亢晓宁（联合培养）、周威（联合培养）、张博（联合培养）、王子昂、夏骥（联合培养）、冯方娜、汪民权（联合培养）、张耀飞（联合培养）、谭舟、付文杰（联合培养）、王继飞、刘可（联合培养）、苏天宇（联合培养）、郭永健、于安然（联合培养）、万鑫、江昊懋（联合培养）、蔡丽杰（联合培养）

Dai Zejun, Zhang Zhe, Gong Meng, Zhu Lei, Wu Yanwen (Co-teaching), Zhang Yiting, Ma Jin, Yin Lu (Co-teaching), Zhong Manlin, Wang Yao (Co-teaching), Meng Ning (Co-teaching), Li Huan, Kang Xiaoning (Co-teaching), Zhou Wei (Co-teaching), Zhang Bo (Co-teaching), Wang Zi'ang, Xia Ji (Co-teaching), Feng Fangna, Wang Minquan (Co-teaching), Zhang Yaofei (Co-teaching), Tan Zhou, Fu Wenjie (Co-teaching), Wang Jifei, Liu Ke (Co-teaching), Su Tianyu (Co-teaching), Guo Yongjian (Co-teaching), Yu Anran (Co-teaching), Wan Xin, Jiang Haomao (Co-teaching), Cai Lijie (Co-teaching)

李兴钢建筑工作室，2020 年 / Atelier Li Xinggang, 2020

摄影 / Photograph:
李欣 /Li Xin

后记
ACKNOWLEDGEMENTS

观、想、做三位一体。这本建筑作品集与同期出版的另外两本书呈现出关联"互引"的关系:《行者图语》是行观之"悟",《胜景几何论稿》则是思考之"述",本书作为实践之"作",系统整理和集成自我的学生时代至今孕育、发生、发展并日益明确的"胜景几何"建筑思考在各个时期行动操作和落地实施的作品,呈现出与时代及现实相对应的建筑师思考和工作实况的丰富性。本书共收纳了 28 个建成作品和一个"装置建筑",作品从内蒙古正蓝旗的元上都遗址博物馆到安徽黄山的绩溪博物馆,从 40 平方米的北京壁园到 14 万平方米的海南国际会展中心,从北京首钢工舍到天津大学新校区综合体育馆,从大院胡同 28 号院改造到唐山"第三空间"综合体,无论地域分布、规模尺度、品种类型、旧城新区,体现出我们的工作所需面对现实的多样性和复杂性,或许实现程度不一,成效与教训并存,甚至语言风格也是各样,但我们始终视真正的建筑(True Architecture)为工作的唯一标准和不变核心,强调和追求实践中思考的某种普适性,希望在实际行动中呈现出思想的价值和关乎未来的方向性探索。翻阅工作室服务器平台上的项目工作区,项目编号自 2003 年至今已列至 153 号,也就是说,我们近 20 年来的 150 余个建筑设计项目中仅落成 30 余项。真正落地实施一个建筑作品无疑是极为艰难的工作,这种艰难当然首先在于建筑与社会、经济、政治、文化、技术乃至时代的密切关联性,但更为切实而关键的艰难性在于,建筑不是建筑师独自可以完成的,需要众多合作者的共同努力——每一个建筑项目背后都有对应的业主(甲方)和施工单位,以及结构、机电、幕墙、照明、交通、经济等专业工程师的参与,同时建筑师自身的工作也必需团队的协同工作,没有这些合作者,建筑无法成为真正的建筑。在此,我要感谢这本作品集中所有项目的业主、施工单位、专业团队以及其他相关的参与和支持方,你们是这些建筑的共同见证者;感谢多年来与我共同战斗的工作室伙伴、同事、学生们,这是一个很长的名单,你们是这些建筑作品的共同创作者;感谢工作室出版小组对所有作品的精心整理、编辑和设计、排版,对我而言,这是一个对过往工作反思总结的有益过程;感谢丁光辉老师的英文翻译和浙江出版联合集团的出版发行,使我们的工作有面向社会和国内外同行展示交流的宝贵机会;感谢程泰宁院士和刘家琨建筑师百忙中为作品集写序,作为我一直敬仰的前辈和兄长,你们给予我最宝贵的鼓励和支持。感谢我的父母、妻儿,你们永远是我忙碌、漂泊之后回归的港湾。

李兴钢
二〇二〇年元月于北京

Envisioning, thinking, and making are the Trinity. This collection of architectural works and the other two books show the relationship of "cross-references": the book of "Wandering, Walking, Viewing, Living" is the "realization" of the travel observations and the book of "Essays about Integrated Geometry and Poetic Scenery" is the "description" of the ideas. As the "work" of practice, this book systematically collates and integrates the works in various periods, as the action and implementation of the " Integrated Geometry and Poetic Scenery " architectural ideas that I have nurtured, occurred, developed and matured since my student days, presenting the richness of the architect's thinking, strategy and work reality in response to the times and reality. This book contains 28 built projects plus 1 "paper building" and 1 "installation building" arranged respectively as the first and the last. From the Xanadu Site Museum in Zhenglanqi, Inner Mongolia, to the Jixi Museum in Huangshan, Anhui, from the 40 square meter Beijing Biyuan Garden to the 140,000 square meter Hainan International Convention and Exhibition Center, from the Silo-Pavilion, Holiday Inn Express Beijing Shougang to the Gymnasium of Tianjin University New Campus, from the Renovation of the No.28 Dayuan Hutong of Beijing to the Tangshan Third Space Complex, regardless of the geographical distributions, size and scale, varieties and types, old or new city areas, they reflect the diversity and complexity of the realities our work needs to face. Perhaps the degree of realization varies. The coexistence of results and lessons, even the language or style is also diverse, but we always regard True Architecture as the only standard and unchanging core of our work. We emphasize and pursue a certain universality of thinking in practice, hoping to present the enduring value of thought and explore the direction that concerns the future. Although covering my 30 years' work as an architect (1991-2020), the book's main works were completed after 2000, that is, after the establishment of Atelier Li Xinggang in 2003. Particularly important works were finished intensively in recent 10 years after 2008. Looking through the project archives stored on the studio's internet server platform, we can find that the project number has reached to 153 since 2003. In other words, only about 30 of the more than 150 architectural projects have been built over the past 20 years. The actual implementation of an architectural work is undoubtedly an extremely difficult task. The difficulty lies in, first of all, the close connections between architecture and society, economy, politics, culture, technology and even the times. Perhaps, the more practical and critical challenge is that a building cannot be completed by the architect alone but requires the joint efforts of many collaborators. Behind every project is the participation of the corresponding client and the construction contractor, as well as structural, electromechanical, MEP, curtain, lighting,transportation, and cost evaluation engineers. At the same time, the architect's own work must also necessitate the team's collaborative efforts. Without these partners, a building cannot be truly built. Here, I want to thank all the clients, construction units, professional teams, and other relevant participants and supporters of the projects collected in this book. They are the common witnesses of these buildings. I thank my partners, colleagues and students, who have worked with me for many years. This is a long list of names; all of them are the co-creators of these architectural works. I thank my studio's publication team for their meticulous work of collation, editing, design, and layout. For me, this is a beneficial process of reflection and summary of past work. Thanks also go to Ding Guanghui's English translation and the publication and distribution efforts of Zhejiang Publishing Group, which offered a valuable opportunity to present our work to national and international audiences. I appreciate very much the forwards written by Academician Cheng Taining and architect Liu Jiakun for whom I have always admired. As a mentor and a brotherly colleague, they have given me the most valuable encouragement and support. Thanks also to my parents, wife and son, who are always there waiting for my return after busy workdays and journeys.

Li Xinggang
Beijing, January 2020

李兴钢，1969 年出生于中国唐山，1991 年和 2012 年分别获得天津大学学士和工学博士学位，2003 年创立中国建筑设计研究院李兴钢建筑工作室，现任中国建筑设计研究院总建筑师、天津大学客座教授 / 博士生导师和清华大学建筑学院设计导师。以"胜景几何"理念为核心，建筑研究与实践聚焦建筑对于自然和人密切交互关系的营造，体现独特的文化厚度和美学感染力。获得的国内外重要建筑奖项包括：亚洲建筑师协会建筑金奖（2019）、ArchDaily 全球年度建筑大奖（2018）、WA 中国建筑奖（2014/2016/2018）、全国优秀工程设计金 / 银奖（2009/2000/2010）等，是中国青年科技奖（2007）和全国工程勘察设计大师荣誉称号（2016）的获得者。作品参加威尼斯国际建筑双年展（2008）等国内外重要展览，并于北京举办"胜景几何"作品个展（2013）。

Li Xinggang received his Doctor of Engineering degree from Tianjin University and founded the Atelier Li Xinggang in 2003. He is visiting professor of Tianjin University and design tutor at the School of Architecture, Tsinghua University. His architectural practice and research focus on the idea of "Integrated Geometry and Poetic Scenery (Sheng Jing)", emphasizing the cultural depth and aesthetic affection in the close reaction of architecture to nature and human-beings. His practice was honored with design awards including the Gold Award of ARCASIA Awards for Architecture (2019), the WA Chinese Architecture Awards (2014, 2016, and 2018) and ArchDaily Building of Year Award (2018), Gold and Silver awards of the National Excellent Engineering Investigation and Design Awards, etc. He received the China Youth Science and Technology Award (2007) and the esteemed National Engineering Survey and Design Master Award (2016). He took part in major architecture exhibitions in China and abroad, including Venice Architecture Biennale (2008), etc., and held his architecture solo exhibition "Integrated Geometry and Poetic Scenery" in Beijing (2013).

出 版 人：徐凤安
责任编辑：王　巍　金慕颜
责任校对：高余朵
责任印刷：汪立峰

装帧设计：姜汶林
李兴钢建筑工作室团队：
　　姜汶林、侯新觉、孔祥惠

图书在版编目（ＣＩＰ）数据

李兴钢：2001—2020 / 李兴钢著. -- 杭州：浙江
摄影出版社，2020.9
　　ISBN 978-7-5514-3028-9

　　Ⅰ．①李… Ⅱ．①李… Ⅲ．①建筑设计－作品集－中
国－现代 Ⅳ．①TU206

　　中国版本图书馆CIP数据核字(2020)第165374号

LI XINGGANG 2001—2020

李兴钢 2001—2020

李兴钢 著

全国百佳图书出版单位
浙江摄影出版社出版发行
　　　　地址：杭州市体育场路 347 号
　　　　邮政编码：310006
　　　　电话：0571-85151082
　　　　网址：www.photo.zjcb.com
经销：全国各地新华书店
印刷：北京卡梅尔彩印厂
开本：635mm × 965mm 1/6
印张：94
版次：2020 年 9 月第 1 版 2020 年 9 月第 1 次印刷
ISBN 978-7-5514-3028-9

定价：398.00 元
本书若有印装质量问题，请与本社发行部联系调换。